STONE

THE DECAY AND REPAIR OF HISTORIC FAÇADES

APW Paye

Published by PAYE

With thanks to Bert, Paul, Peter, other directors and staff for their interest and care,
to Viv for pulling this together, to Anna for support, and to Felicity and Alice for forgoing time.

Published by PAYE
Copyright © APW Paye
www.PAYE.net

ISBN 978 0 9930904 0 0

By Appointment to
Her Majesty The Queen
Building Facade Restoration
and Conservation

STONE

THE DECAY AND REPAIR OF HISTORIC FAÇADES

APW Paye

Published by PAYE

FOREWORD

Those of us who have been lucky enough to have employed PAYE as contractors have conceived a great admiration for them as a family firm with a great tradition of love for their craft. Their knowledge of stone, of how to work it and of the history of stone masonry in this country, informs their approach to their business and the infectious enthusiasm for what they do makes them a pleasure to work with.

This book will be an invaluable companion for anyone who owns or manages buildings in this country. As you leaf through it, you will recognise many of the subjects of the photographs that appear in it and suddenly understand what an important part PAYE have played in fashioning and repairing the buildings that set the tone of our lives.

Let us hope that they and the craftsmen they employ will be about for many decades to come.

The Marquess of Salisbury, KCVO, PC, DL
May 2014

Hatfield House, Hatfield,
Hertfordshire, AL9 5NF

CONTENTS

FOREWORD	03
INTRODUCTION	07
1 STONE TYPES & CHARACTERISTICS	09
Limestones	09
Sandstones	11
Marbles	12
Slates	14
Granites	14
2 DEFECTS	17
Definition of a Defect	17
Weathering	17
Failure of Cement Mortar Repairs	18
Salt Damage (crypto-florescence)	21
Clinker Concrete	22
Corrosion of Embedded Metalwork	25
Inappropriate Surface Treatments	27
Rainwater Management	28
Frost Damage	31
Vegetation	31
Cornices	32
Pediments	34

3 CLEANING	39
Historic Context	39
Selecting the Most Appropriate Method	41
The Various Methods	45
High Pressure Water Lance	45
Fine Nebulous Water Sprays	47
Chemical Cleaning	49
Pressurised Hot Water/Steam	50
Grit Blasting	53
Low Pressure Air Abrasion	53
Poulticing	56
Method Reference Guide	59
Sequencing	59
Final Clean	60
4 REPAIRS	63
Planned Maintenance Periods	63
Surveys and Identifying the Causes of Failure	67
Rainwater Management	68
Ordinary Portland Cement (OPC)	68
Repointing	70
Redressing (Reworking) Stone	75
Replacing Stone	75
Mortar Repairs	78
Sacrificial Shelter Coats	81
Stitching Fractures	81
Consolidation of Friable Surfaces	82

5 TECHNICAL ISSUES, MANAGEMENT & ORGANISATION 91
Strength in Compression 91
Orientation of Bedding Planes 91
Selection 93
Scheduling / References / Drawings 94
Unloading, Distribution and Installation 99
Personal Protection Equipment (PPE) 99
Scaffolding 101
Lifting Equipment 105
Tolerances 106
Shims 109
Block Bonding 109
Grouting 109
Movement Joints in Traditional Masonry 109
Thin Cladding Systems 110
Casting Concrete Against Masonry 116
Bedding Cills 116
Cantilevered Stairs 118
Protecting a Retained Façade After Demolition 123
Managing Differential Movement Between a Retained Façade and New Structure 123
Dismantling, Storing and Reconstructing Historic Façades 127
Historic Use of Metals 133
Corroding Steelwork 135
Surveys 137
Steel Treatments 138
Replacement Masonry 141
Maintenance Strategies for Steel Framed Façades 141
Cathodic Protection 143

6 ALTERNATIVE MATERIALS
Reconstituted Stone 153
Brickwork 155
Brief History 156
   Manufacture 160
   Classifications 159
   Bonding 159
   Defects 161
   Repair 163
Terra Cotta and Faience 165
Stucco Render 171
   Brief History 171
   Defects 174
   Repair 174

7 A STONEMASON'S HISTORY OF LONDON 179
The Founding of London 179
After the Romans 182
The Impact of the Norman Invasion 184
The Early Modern Period 195
After the Great Fire of 1666 195
The Post-Industrial Period 196
The Industry Today 200

GLOSSARY 204

INDEX 212

INTRODUCTION

This book is intended as a guide to best current practice for those responsible for the conservation, maintenance and repair of historic masonry and façades. It is not intended as a guide to the craft of stonemasonry: many other books have already covered this in great depth.

The word 'current' is used to illustrate the fact that what one generation once deemed to be best practice has often been found subsequently to be harmful. Examples include types of mortar, cleaning chemicals, and the use of inferior materials. I was taught in the 1970's that a good mortar was a strong mortar, typically three parts of well-graded sand to one part of Ordinary Portland Cement. We now know that these and other similar mortars, if used to bed, re-point or 'restore' a natural limestone, will cause that stone to deteriorate. Unfortunately, the use of these mortars increased after the Great War, extensively so after WW2, so now their legacy is of failing stone on the façades of many of our important historic buildings. Regrettably, we still see these mortars being specified by professional consultants and often used by our competitors.

Decay caused by the use of inappropriate techniques and materials can occur within seconds, minutes, and hours, though usually it occurs over months or years. Often the result of bad practice is not realised nor fully understood for decades. It would therefore be arrogant to believe that our current techniques are infallible and we should recognise that our knowledge must evolve as we forever seek to better maintain our national heritage.

The notes and guidelines set out in this book are our opinions and, of course, other opinions might be equally valid. Technical details, problems and appropriate solutions will vary from one project to another and often the most successful projects are those where knowledge and experience is shared between the skilled experienced artisans and the professional consultants. Through such collaboration, problems can be properly identified and the best repair strategies developed and implemented.

In the following pages we have attempted to share the knowledge that has been accumulated by the many individuals at PAYE over the many decades that we have spent repairing and conserving historically important masonry façades.

It is not possible to provide answers to all questions but we hope that we have covered the important principles to help ensure that yesterday's mistakes are not repeated tomorrow.

PAYE have undertaken conservation work, repaired, cleaned and adapted all the buildings featured.

CHAPTER ONE

STONE TYPES & CHARACTERISTICS

In geological terms there are many different types of stones but in our context there are only five generic categories that usually concern us: limestones, sandstones, marbles, slates and granites. Most stones used in the construction of masonry façades will fall into one of these categories. Artificial stones such as terra cotta, faience, reconstituted stone and stucco render are described in *Chapter 6*.

Different building stones react to pollution, rainwater saturation, cleaning techniques, and inappropriate surface treatments in different ways. The following descriptions identify the characteristics of each of the generic categories.

LIMESTONES

Limestones are sedimentary stones formed in beds under warm shallow seas and laid down over millions of years as creatures died and sank. The bones, teeth, shells and claws of these creatures provided the calcium carbonate that makes up about 97% of a limestone. Look closely at a limestone and you will see fossils, some

Weathered Portland stone finial details at St Paul's Church, Mill Hill

whole and others fragmented. Examples of limestones used in London and the south-east of England include Portland, Bath, Kentish Ragstone, Caen, Ancaster, Clipsham, Hopton Wood, and Clunch.

Sedimentary stones can be described as a natural concrete being comprised of aggregates and binding cements. In a limestone the aggregate is formed of fragmented fossil shells, bones, teeth, claws and particles of calcium carbonate. The binding cement is the calcium carbonate matrix which binds the aggregate together. Portland, Caen, Ancaster, and Bath stones are sometimes referred to as oolitic limestones because the calcium carbonate particles were once thought to be tiny eggs or fish roe ('oolith' means eggstone).

Limestones, if clean, tend to be light in colour (our own bones and teeth are an off-white colour). The colours of a limestone will vary with differing amounts of iron and other minerals.

Many limestones can be polished by cutting and abrading the surface with fine polishing tools. However, a polished limestone surface will not last long as such if used externally because rainwater will slowly dissolve away the binding matrix causing the surface to erode. The more

STONE TYPES & CHARACTERISTICS

Portland and Bath stone on the façades of the Corinthia Hotel, converted from government offices

polluted the atmosphere, the more acidic the rainwater will be and the quicker the limestone surface will erode. The 1956 Clean Air Act banned the burning of coal in city centres and from this time pollution and the rate of erosion improved from a typical 1mm every 10 years to 1mm every 100 years.

If a limestone is to be used externally it is usually finished with a fine rubbed honed finish.

SANDSTONES

Sandstones, like limestones, are formed as sediments. They are formed in layers referred to as bedding planes (often planes of weakness) from sand particles washed off eroding mountains into rivers and estuaries, and cemented together by different types of binding matrixes: calcium carbonate, clay, iron, and silica or various combinations thereof. Examples historically used in London and the south-east of England include York, Bramley Fall and Stanton Moor. In Kent and Sussex, Wealden stone has traditionally been used.

Sandstones tend to be more intensively coloured than limestones. Sandstone colours are determined by the colour of the original sand, the type of binding matrix, and other minerals present. Reds and buffs are common.

Different types of sandstones have very different weathering characteristics – some are much more durable than others. Clay is not a durable binding matrix material in the UK's wet environment so sandstones containing a high clay content are not suitable for building façades.

Some sandstones (such as York stone) can be readily riven (ie. split) along their bedding planes and used as paving stones

STONE TYPES & CHARACTERISTICS

ABOVE AND RIGHT
Carved Carrara marble indent repairs on the statue of Shakespeare, Leicester Square

or roofing tiles. A riven finish would historically have been much quicker and cheaper than working the stone to a smooth finish by hand. Sandstones are less susceptible to surface erosion from acid rain than limestones.

MARBLES

Marbles are limestones that have been buried and subjected to extreme pressure and heat for millions of years. In consequence their properties have changed, and they are therefore described as metamorphic stones.

Marbles can be polished and colours may be vivid – commonly reds, pinks, greys, greens, yellows, white or black. The term 'marble' is often mistakenly used to describe any limestone that can be polished.

There are hundreds of different marbles, many of which were traditionally sourced from Italy, Spain, and Portugal. A famous example is a white veined marble from the Carrara mountains of Tuscany which has been quarried continuously for thousands of years. This is available in a variety of grades and traditionally used for statuary, internal paving and wall linings. Famous English marbles are Ashburton, which is no longer commercially available, and Purbeck. Both these stones are technically limestones that can be polished rather than marbles.

Marbles are rarely used in the construction of external façades due to the relatively high cost of transporting materials over long distances, the relatively higher cost of working the relatively harder stone, and the fact that any polished surface would quickly erode in polluted city atmospheres. There are two examples of buildings with solid marble stone façades in London's Piccadilly on opposite sides of St James's Street.

New carved Carrara marble angel, private chapel, Hatfield House

STONE TYPES & CHARACTERISTICS

ABOVE
Henry Moore Arch in Travertine marble under construction, Hyde Park

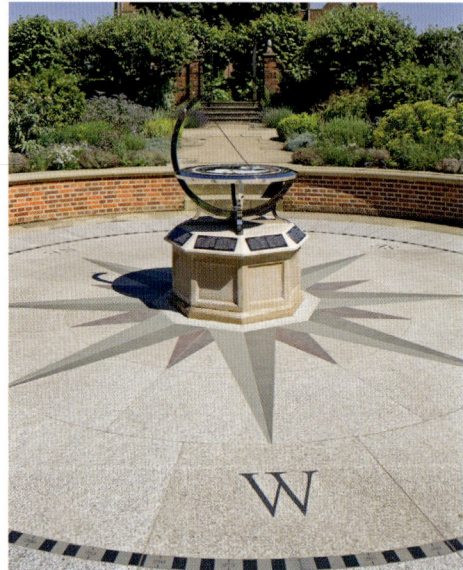

RIGHT
UK sourced granite, slate, and Clipsham stone used to construct the Longitudinal Dial, Hatfield House

SLATES

Slates are also metamorphic stones, having originally been sedimentary muds and clays that have been subjected to extreme pressure and heat for millions of years.

There are good sources of slate in the UK. Quarries in Cumbria provide green and grey varieties; and various Welsh quarries produce purple (Penrhyn), blue (Cwt-y-Bugail), green (Nantlle), grey (Llechwedd), and black (Corris).

Slates are durable materials which were traditionally riven and widely used for roof tiles, or used locally as walling stones and paving. They can be polished, though as with marbles and limestones, the polished finish will erode and fade if used externally.

GRANITES

Granites are very durable, appreciably more dense than other building stones and have greater impermeability. They are also much harder and are therefore more difficult to work. This makes them expensive compared to limestones and sandstones for anything other than machine cut thin plain areas.

Granites can be polished and, unlike marbles, the polished surfaces will usually endure externally in polluted city atmospheres.

Grey granites from Cornwall were commonly used on historic façades in London traditionally as plinths at pavement level. Some are currently commercially available.

Other traditional sources of granite were Scotland, Finland and Norway. Colours used extensively in 19th century London tended to be greys, pinks, reds and black. Few Scottish granites are currently commercially available.

Perhaps surprisingly, the UK was the world's largest exporter of granites in the 19th century. Today, the cheapest source of granite (other than in terms of its carbon footprint) is China. Italy, India, Greece, South America and Spain are other sources. Most granite today is imported as polished slabs and cut to size in the UK. Three dimensional solid blocks of granite are usually special orders cut and worked to finished dimensions overseas.

Colours can be as varied as marbles.

Kentish Ragstone and Bath stone indents on the façades of the converted St Georges Church, Tufnell Park

CHAPTER TWO

DEFECTS

The key to a successful conservation programme is an accurate understanding of the causes of failure. Only then can a suitable repair strategy be developed to enhance the long term and cost effective performance of an historic façade. This chapter describes the most commonly encountered defects, but first discusses when a defect should be considered as such.

DEFINITION OF A DEFECT

Failure is subjective and its definition can vary from one project to another. In the 1970's and 1980's, a nominal weathering loss of profile would be deemed to be failure and the stone would either be replaced or repaired with (often inappropriate) mortars. Today a nominal loss of profile is generally not regarded as failure unless it has occurred to the extent where support of the stone above or protection of the stone below is compromised. Aesthetics also play a part - weathered stones might be deemed to have 'failed' because they are perceived as being unsightly. It is important to understand the client's requirements because, costs are proportional to the amount of repairs undertaken.

On historic buildings, the retention

Wealden stone indent repairs, Tonbridge Castle

and conservation of original material is an important issue which is often considered the objective. Masonry which might be deemed to have 'failed' needing replacement on one project might be highly valued and deemed suitable for conservation on another.

Most clients wish their façades to appear well maintained though often an owner's requirement for an immaculate façade on their corporate head office contradicts a conservation officer's requirement for a minimal amount of cleaning and repair, and we often find ourselves in a position where we are tasked with 'brokering' a pragmatic compromise which not only satisfies both parties but does so within a limited budget. Thankfully we have enjoyed considerable success in achieving this.

WEATHERING

Weathering is the term used to describe surface erosion caused by the actions of wind and rain, and often exacerbated by the presence of atmospheric pollutants that exist as deposits on the masonry surface or in solution in rainwater.

The Clean Air Act 1956 stopped the burning of coal in city centres and this dramatically reduced the levels of pollution. Before this time, smoke and fog would combine to form thick dense clouds referred to as 'pea soupers' or 'smogs' where the atmosphere had a similar acidity to that of lemon juice. Not only was this harmful to city dwellers themselves (tens of thousands of Londoners died during the smogs), but it was extremely damaging to the limestone buildings. Much of the weathering loss of profile that has occurred on masonry surfaces in city centres dates from before the 1956 Act.

Limestones dissolve in rainwater. The more polluted and acidic the rainwater, the quicker the limestone will dissolve. The rate that an averagely durable limestone dissolved in polluted London was approximately 1mm every ten years. Now the air is cleaner, the current rate is approximately 1mm every hundred years. As the binding matrix of the limestone dissolves away, the fossil fragments gradually become more exposed. Stones in susceptible locations on buildings constructed in the mid-19th century are often found to have 10mm or more of their fossil fragments projecting. In the 1970's and 1980's it was common to be instructed to carefully cut these off as they were deemed unsightly. This is not considered good practice today.

Another type of pollution damage was caused by the sulphates and carbonates from the burning of coal reacting with the binding matrix of the limestone. This reaction occurred on the surface of the stone and formed a thick black crust. As the binding matrix was drawn to the masonry surface to form the thick crust, the stone immediately beneath the skin became depleted of its binder and therefore became weak and friable. Expansion of the depleted zone also occurred which tended to blister off the formed crust thus making the stone substrate very susceptible to wind and rain erosion.

FAILURE OF CEMENT MORTAR REPAIRS

In the 1970's and 1980's it was common for the profiles of weathered stone to be reinstated using cement mortars. Cement mortars are generally less permeable than limestones and so tend to trap moisture. Raised levels of water retention cause the limestone to deteriorate locally with the result that the bond adhering the cement mortar becomes unstable. The time taken for the repair to fail will vary with the relative differential in permeability between the stone and the mortar, and the degree of exposure of the stone to repeated saturation from rainwater. Mortar repairs on copings and cills will fail before similar repairs on protected ashlars.

Many of the relatively porous Bath stones fail where cement mortars have been used. Bath stones were popular in London from the early 19th century (economically transported to London after the opening of the Kennet and Avon canal linking the rivers Avon and Thames) and most of the façades constructed at this time have not only weathered considerably but have been 'repaired' using cement mortars. These repairs are known to fail within a decade or two.

It is difficult to now understand why it was common practice to 'restore' the original profiles of eroded stone buildings. It was probably believed to be appropriate to attempt to make the façade appear new. Today, we accept that an older façade should appear as such. It is regrettable, however, that some companies still undertake such inappropriate work today.

RIGHT
Failing cement mortar repairs on terracotta

DEFECTS

ABOVE
Weathered Chilmark coping stones, Cranborne Manor

LEFT
Defective Bath stone balustrading salvaged from the burnt ruins of an earlier building and decorative gauged brickwork revealed after the removal of defective render at Cliveden House, Buckinghamshire

SALT DAMAGE (CRYPTO-FLORESCENCE)

Contaminating salts can be present within the masonry and dormant for many years. Changes in moisture levels or temperatures can be enough to mobilise these salts and cause crystallisation damage. This can be relevant when an historic building is renovated and heated after years of neglect.

Marine salt is another source of harmful contaminant. Façades within 3 miles of a coastline often become contaminated by salt as the sea spray is blown on shore. The stone surfaces absorb the saline solution and the salt crystals form where the moisture evaporates. This is usually at a depth of 1 or 2mm below the stone surface and is sometimes referred to as crypto-florescence. Where the crystals form in a stone with a high proportion of micro pores, the stone will be more susceptible to damage because the crystals grow larger than the pores and consequently spall the surface. Stones with a greater proportion of large pores tend to have a greater resistance to salt crystallisation and are thus more durable.

Pavement de-icing salts are another common source of contamination as are any harmful salts present in ground water.

DEFECTS

ABOVE
Pollution damage on Bath stone and brickwork

RIGHT
Marine salt crystallisation damage on Caen stone, Anglican Cathedral, Bermuda

Salt damage typically forms cavities on the surface of the stone which deepen as the damage continues.

Limestones are regarded as being more susceptible to salt damage than other types of stone because the calcium carbonate reacts with acidic atmospheric pollutants to form sulphates which crystallise on or near the masonry surface.

It is considered best practice to draw contaminating salts out of the masonry (a process known as desalination) using poultices. Attempts to seal the masonry surfaces will exacerbate problems because moisture will eventually find somewhere from which to evaporate, and in these locations salt damage will be excessive. Trapped moisture will also damage the stone. For this reason, surface treatments, particularly those that attempt to seal the surface, are best avoided.

Preventing the stone absorbing harmful salts is difficult in practice. Façades in marine environments will always be susceptible to salt damage, as will masonry absorbing harmful salts from the ground. Sacrificial shelter coats can be used to protect stone in marine environments.

Controlling damage caused by the absorption of contaminated ground water will be difficult. Introducing a dpc might help to protect the masonry above, but problems below dpc level will be exacerbated because this is where the moisture, salt contamination, and evaporation will be concentrated. The best way to protect a façade from damage caused by the absorption of salts from the ground is to construct that façade up to a level above ground using a dense, impermeable material (for instance certain types of granite), though doing this retrospectively will probably be on the impossible side of expensive and unlikely to gain the approval of the local conservation officer.

CLINKER CONCRETE

The aggregates present in a clinker concrete contain a wide range of materials such as part-burnt coal, burnt shale, carbonaceous ash and slag. Clinker concretes were typically used as a cheap void filling material around steel floor joists and rafters.

The aggregates contain potentially harmful sulphates which form sulphuric acid if water is allowed to percolate through the clinker concrete. Sulphuric acid is an effective electrolyte which will encourage any embedded steel to corrode.

Another problem with clinker concrete is that part-burnt coal is dynamically unstable and will expand when wetted. We have known 225mm thick granite column sections to be sheared by the expanding action of clinker concrete.

Clinker concrete was often used to fill hollow terra cotta blocks before installation. It is therefore not uncommon to find terra cotta coping, cill and cornice units fractured as a result of the clinker fill expanding where the units become saturated by rain.

DEFECTS

CORROSION OF EMBEDDED METALWORK: CRAMPS, CHAINS AND SPIDERS

Corroding steel frames are discussed further in *Chapter 5*.

Loadbearing masonry walls built before the early 20th century were typically constructed with ferrous cramps set across perpend joints securing adjacent stones together. The cramps often do not seem to have any particular structural function except to hold the units together during the long curing periods of the lime mortars. Occasionally cramps are found to have other specific structural restraint functions.

Cramps are generally fixed into the upper face of adjacent stones across their perpend joints, and were often set in molten lead as an early form of galvanic protection. If a cramp corrodes, it expands in volume and this results in the stressing and fracturing of the embedding stones in characteristic U-shaped spalls, sometimes occurring only on one side of joint. When surveying façades built before the 20th century, this pattern of spalling will often be found to be the most apparent defect.

When façades were later constructed with embedded steel frames (approximately 1906 to 1937), cramps were often used to restrain cornice, fascia, and soffit stones to the steel frame.

Cramps were also used to tie courses of domes and spires together in order to restrain them against spreading under the imposed loads of the masonry above. In these conditions, the cramps will be performing a permanent structural function and must be maintained or carefully replaced. From the 17th century, iron chains were sometimes embedded into the masonry to perform this lateral restraining function (as at St Paul's Cathedral).

It is also common to find a series of restraint ties embedded across the centre of spires to stabilize specific courses of masonry. It would be usual to find two or three levels of such restraints at approximate third or quarter heights of the spire. Ties built into each face of a spire and connected at the centre are referred to as 'spiders'.

Another common feature is the requirement to stabilize the top of a spire which, because it is not loaded by the mass of masonry above, can become unstable under wind loads. The top apex stone is often tied down to the 'spider' below. Another means to tie down the apex stone of a spire is to suspend from it a large rough stone by means of a chain.

Weather vanes above a spire are often fixed to a central rod which extends down and is secured to the first spider below.

Ferrous metals will corrode, even if bedded in molten lead for galvanizing protection (as was specified by Wren at

LEFT
Terra cotta blocks damaged by clinker concrete fill, Victoria Quarter, Leeds

ABOVE
Spall caused by corroding cramp in 19th century Wealden stone, Pembury, Kent

St Paul's Cathedral). When such metals corrode, they will expand by up to thirteen times their original volume and in so doing will stress and fracture the surrounding stone.

The closer the embedded metalwork is fixed to the surface of the stone and the more exposed that masonry surface is to repeated or prolonged saturation from rainwater, the more extensive will be the corrosion.

The generally accepted repair strategy is to carefully expose every corroding cramp and to either treat, replace or remove it. The decision as to which is the most appropriate action will be a case of judgment on an individual basis. Unless the cramps themselves are of historic significance, the most appropriate method of replacement will be to install a stainless steel threaded dowel set in hydraulic lime mortar or resin, if deemed necessary.

The affected stones are usually repaired by indenting new sections of matching stone either side of the joint (ie two indents). It is important that the joint is maintained: a single indent installed across the joint might be easier but will eventually fracture and then allow more rainwater to penetrate.

It is usually considered inappropriate to repair, treat or replace cramps that are not corroding on the basis that if they were going to corrode they probably would have done so by now. 'If it isn't broken, don't fix it' is the maxim which generally applies.

It is very important that rainwater is properly managed. Other than where the cramps have been installed too close to the surface, most defects are caused by poor rainwater management.

INAPPROPRIATE SURFACE TREATMENTS

The failure of masonry façades to withstand the degrading effects of weather and pollution has encouraged many individuals and corporations over the last 150 years to attempt to develop miraculous surface treatments which purport to protect the stone from all known ailments. Some of the consequences of these 'remedies' have been disastrous. In particular, the sadly common use of silicon treatments in the 1970's resulted in many façades losing several millimeters of delaminated stone where the treated stone peeled away from the untreated stone substrate. This usually occurred within a period of approximately 10 years.

The cause of delamination is believed to relate to the entrapment of moisture below the treated surfaces where increased local levels of moisture cause the stone substrate to degrade. Another contributory effect

LEFT
Hawksmoore's Portland stone St Alfege Church, Greenwich

ABOVE
Exposed corroding steel cornice support system on the faience façades of the Regents Palace Hotel, Regent Street

DEFECTS

ABOVE
Corroding steel beam ends and clinker concrete in Red Mansfield stone façades, Hippodrome, London

RIGHT
Refurbished bronze balustrade and Carrara marble, Leicester Square

could be the possible variance in thermal capacity between the treated and untreated layers of stone. Both effects could cause a plane of weakness to be created parallel to the surface where the delamination occurs.

Today's treatments appear to be better. We have known façades to be treated with siloxanes and not suffer any delamination after a period of 30 years. However, because surface treatments are designed to be absorbed deeply into the stone pore structure, they are not reversible and this permanence does not fit well with conservation philosophy.

Our advice is therefore to generally recommend against the use of any surface treatment. However, in exceptional circumstances, it might be appropriate for a consolidant to be applied to individual masonry units where severe surface erosion is occurring but the client does not wish for the stone to be replaced during an ongoing phase of repair work. Consolidants are discussed in further detail below.

RAINWATER MANAGEMENT

Many (if not most) defects are caused by repeated or prolonged rainwater saturation of the masonry. Therefore, the effective design and maintenance of the rainwater collection and disposal system is the most important consideration for developing an effective conservation repair strategy.

Typical problems caused by leaking rainwater goods are staining, efflorescence, decay, and corrosion of embedded steel fixings, beams, and columns. The cost of rectifying these defects can be considerable.

Most large cornices were originally designed for the rainwater to run back into a gutter in the sky surface which was connected to an internal rainwater pipe by means of a small lead 'swan-neck' pipe. This is a poor detail because the connecting lead pipe soon blocks with debris (usually an unfortunate pigeon) and rainwater ponding on the cornice eventually finds its way

through the cornice to stain and erode the masonry below or, worse, cause embedded steelwork to corrode. If attempts are made to use drain clearing rods to remove the debris, the likelihood is that the pipe will become damaged and the water will soon again find its way through the cornice to cause damage below.

An improved detail will involve installing a protective lead capping (usually Code 5 or 6) over the cornice (making the gutter and sawn-neck lead pipe redundant) and adjusting the falls so that rainwater drips off the leading edge of the cornice. The leadwork must be installed to Lead Sheet Association guidelines, and the drip detail over the leading edge looks best if it is installed as discretely as possible.

Any concentration of rainwater on a stone façade has the potential to cause erosion, staining, water ingress, and the corrosion of embedded steel.

Water running off glass will saturate the stone below, requiring a properly designed cill detail with falls, drips, throats, stoolings, waterbars, and dpc's to avoid problems. Coping stones need to be similarly designed.

FROST DAMAGE

Ice crystals forming and expanding in the masonry pore structure can cause physical damage on the surface of the stone. However, most building stones have the ability to allow ice to form near the surface without physical damage. Good detailing of the masonry to prevent repeated or prolonged saturation and the installation and maintenance of an effective rainwater disposal system will help further.

VEGETATION

Shrubs growing within open masonry joints will have roots that can grow to excessive sizes, displacing and fracturing masonry as it does so. If left unchecked, the only remedy will often be to carefully dismantle and reconstruct the masonry, removing the roots in the process. Regular maintenance involving the pointing of all open joints will reduce the risk of shrubs becoming established.

Façades should always be regularly inspected and shrubs removed as soon as possible. If it is not possible to fully remove all roots nor dismantle the masonry surrounding them, the shrub must first be treated with an appropriate biocide before the accessible part of the shrub is removed. Further treatment of the masonry with an appropriate biocide is also recommended before the area is covered by any subsequent repair.

FAR LEFT
Damage caused by inadequate rainwater management at the Elizabeth Garrett Anderson Hospital (now Unison House), Euston Road

ABOVE
Saturated Portland stone stained by the migration of soluble organic residues.

DEFECTS

ABOVE
Moss and lichen growing on the Portland stone Eleanor Cross at Charing Cross Station

ABOVE RIGHT
Moss growing on slate roofs

RIGNT
Completed conservation work on stucco detailing including retained GRP, GRC, timber and Jesmonite repairs on the façades of Kenwood House

Mosses and lichens growing on historic masonry are sometimes considered important, particularly in Special Sites of Scientific Interest where we have successfully dismantled and reconstructed historic masonry ensuring that the rare mosses and lichens were carefully preserved by always maintaining the original orientation of the masonry towards sunlight during the whole process.

Some algae and lichens secrete acids which can superficially harm and stain building stones.

CORNICES

In constructing traditional loadbearing façades, the mason's rule of thumb installing cornices was that never more than one third of the cornice's total width should be unsupported, in other words, two-thirds of the cornice should be supported by the masonry façade. The centre of gravity of the cornice stones was therefore always back inside the façade wall line, and the mass of masonry above would keep the overall centre of gravity of the complete façade close enough to the overall centre of the wall.

Over the years, steel framed façades became lighter, façades became thinner, and window areas increased. The use of heavy stone features became more and more daring and the 'two-thirds supported' rule was being pushed often to the point where large cantilevered cornices could not be stabilised by the masonry below nor the mass of masonry above. They needed to be supported, restrained and strapped down to the embedded steel frame in order to keep them stable. This leads to problems on the external returns of these cornices where the stones are necessary large – we have known them to weigh up to 4 tonnes each – and the method of stabilising has been inadequate. The often alarming result is that these stones, cantilevering on two sides, gradually rotate out of position, requiring the careful installation of additional support and restraint fixings.

DEFECTS

Traditionally, cornices were designed to collect rainwater and often featured a gutter in their sky surfaces. These gutters either connected to rainwater pipes or larger gutters behind parapets by means of small lead pipes or chutes. These are very prone to blocking and susceptible to damage if rodded. For this reason it is considered good practice to adapt the falls on these cornices so that rainwater is diverted away from the façade to drip down to the pavement below. As it will only drip during rainfall, inconvenience to pedestrians below is only marginally increased. The cornice should be protected in a suitable material, usually Code 5 (or 6) lead, with suitable discrete drips, upstands, falls and cover flashings.

PEDIMENTS

Large pediments constructed over façades without the mass of masonry above can suffer long term rotation and displacement due to their centre of gravity eccentrically loading the wall below. Problems become apparent on any returns and manifest as vertical fractures. It is usually possible to stabilise the pediment masonry by installing restraining straps tying the pediment stones back to the roof structure behind. It is necessary to consult a structural engineer in designing the strapping arrangement to ensure that the restraining stresses can be borne by the adjacent structure.

ABOVE
The English Heritage craftsmanship award winning restoration of Hadlow Tower

RIGHT
Pediment detail of the repaired north portico, British Museum

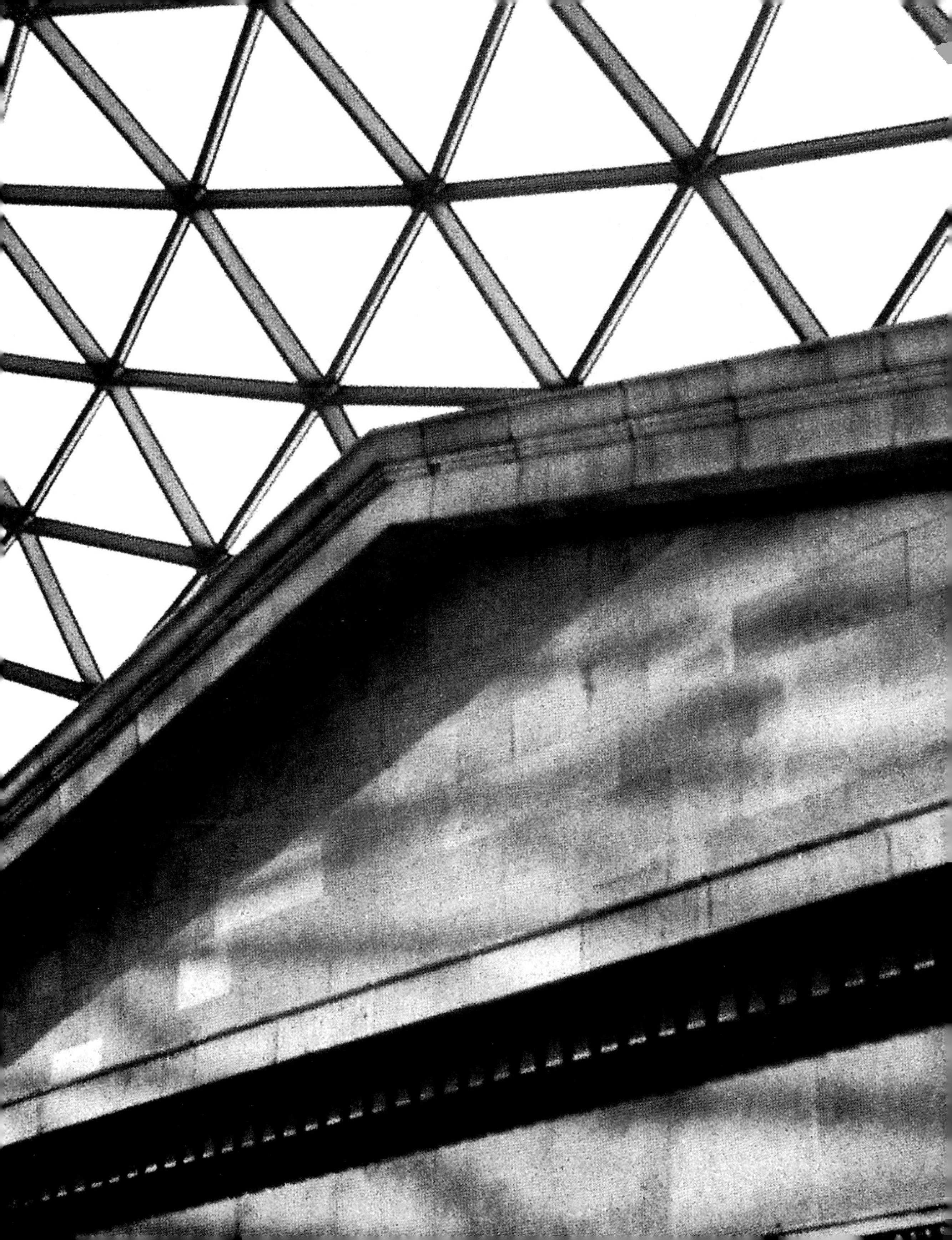

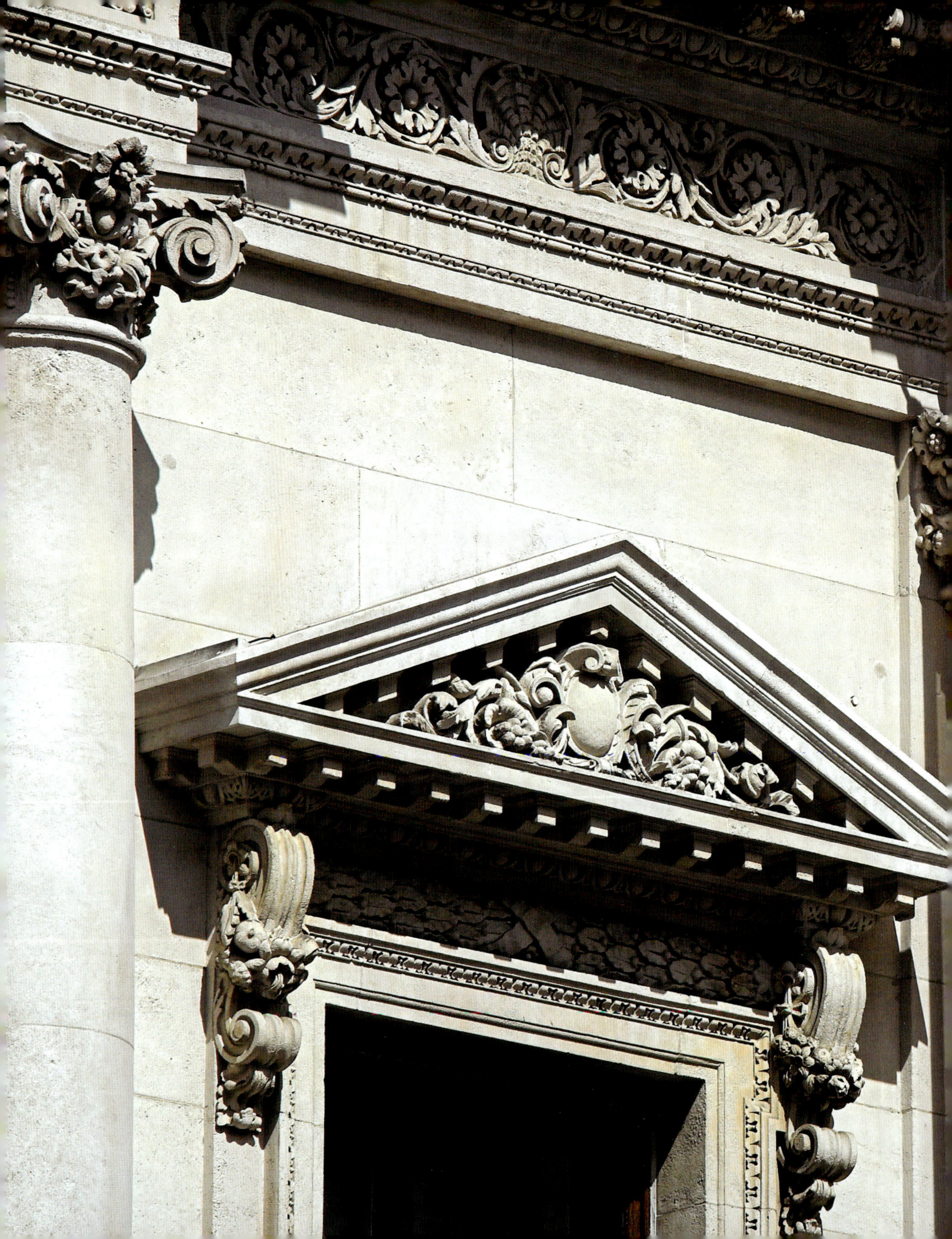

Cleaned and repaired Portland Stone at Burlington House, Piccadilly

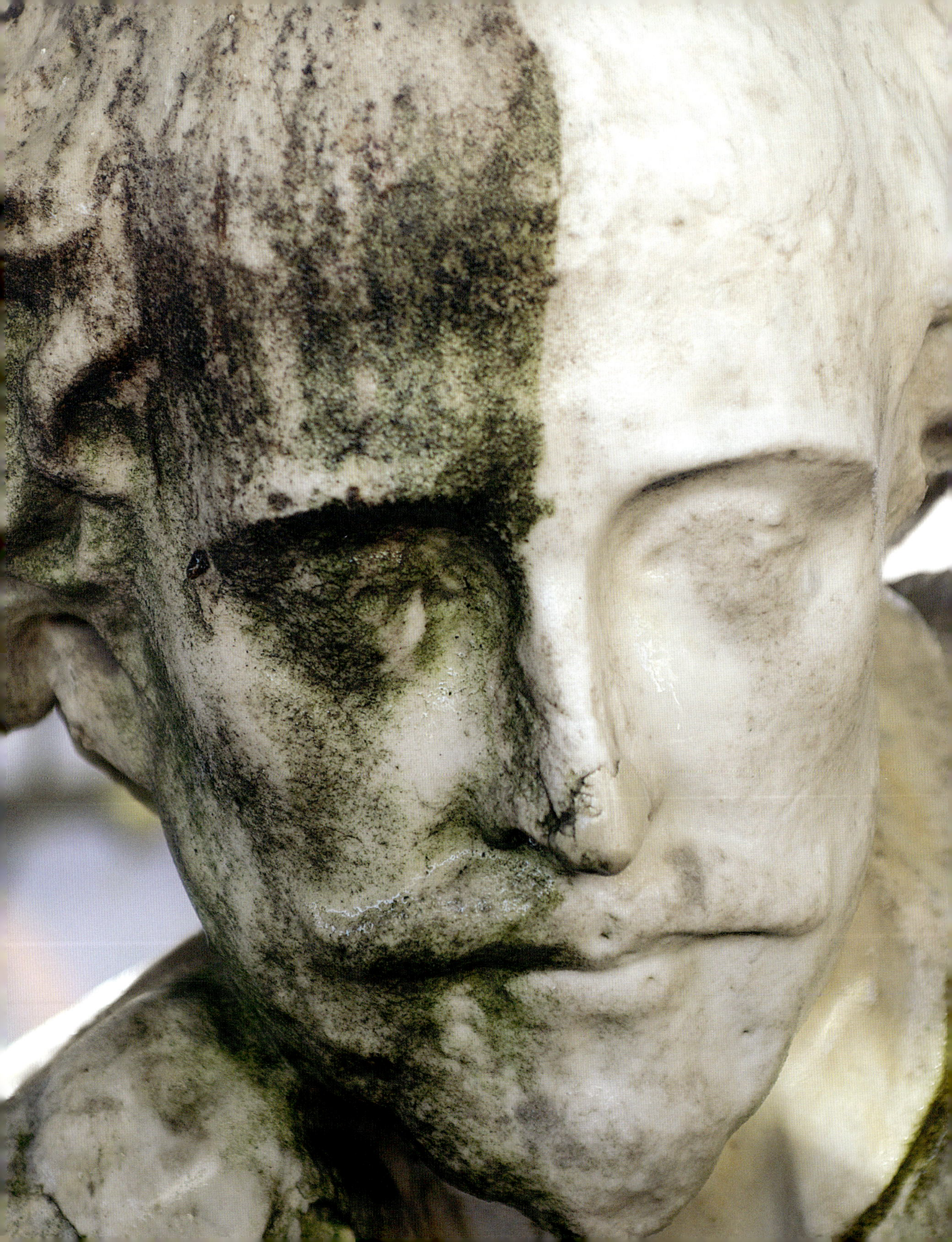

CHAPTER THREE

CLEANING

It could be considered that cleaning a façade has the greatest visual impact of any work undertaken upon an historic building. This appeals to many clients but it comes with risks because cleaning a building inevitably results in nominal surface change or damage no matter how much care is taken. Usually such damage is absolutely minimal, but the risks are real. Cleaning a façade therefore is a three way balance between desired appearance, the risk of surface damage, and cost. Costs are relevant because cheaper methods tend to result in more extensive damage.

Cleaning methods involving the use of abrasives and chemicals have the greatest potential to cause damage. High abrasive pressures and inappropriately hard and large abrasive particles will make a façade appear clean at relatively low cost but will severely damage the surface, resulting in the façade quickly re-soiling as dirt particles collect more quickly on the newly abraded surface. Such damage is a criminal offence.

Carrara marble statue of Shakespeare undergoing careful cleaning, Leicester Square

HISTORICAL CONTEXT

The practice of cleaning façades commenced in the 1920's initially using methods such as sand-blasting and steam cleaning. City centre façades used to be black with soot from the burning of coal. I've been told by people living in London at that time that seeing a cleaned building for the first time had a large impact as not many people could recall the original appearance of the building nor the colours of brick and stone beneath the soot.

The use of sand as an abrasive was dangerous for the operatives due to the increased risk of developing silicosis. Operatives gained some protection by using helmets fed with clean air from the compressor however the dust would have been as dangerous for the public as well as anyone tasked with the responsibility to collect and dispose of it. The use of sand gave way to other abrasives such as aluminium oxides, olivine, and calcite.

Steam cleaning was also soon recognized to be damaging due to the fact that sodium bicarbonate was added to the water to control limescale. This had a detrimental effect on the newly exposed stone surfaces because efflorescence caused by the crystallisation of salts within the masonry pore structures physically damaged the stone surfaces. Experiments undertaken

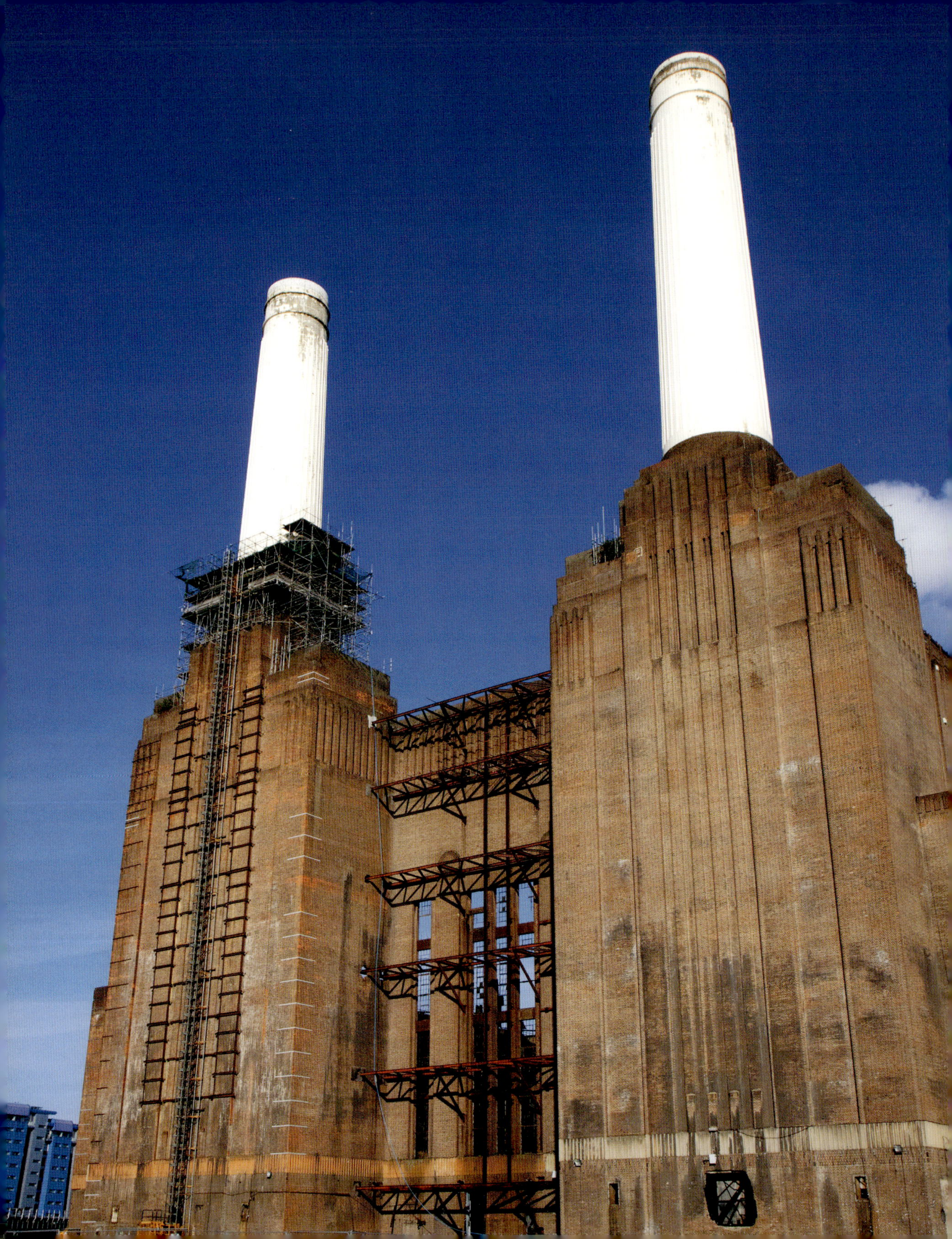

CLEANING

in 1932 showed decay becoming apparent within 12 months of cleaning. Masonry cleaning companies at the time would be requested to provide guarantees that such additives would not be used.

The use of chemicals increased in the latter part of the 20th century. Again, problems were experienced and façades permanently damaged, some severely. Hydrofluoric acid is sometimes used to clean brickwork and terra cotta but is known to have detrimental effect on limestones and many sandstones. Technical data sheets of the 1980's stated that certain products containing hydrofluoric acid could be used to clean any type of natural stone. It is impossible for any single product to do this due to the varying chemical compositions of the many different stones, but similar statements can still be read on containers of cleaning chemicals in DIY stores today.

History has taught us that new cleaning techniques which are not fully tested are potentially damaging. Therefore, caution must always be exercised particularly as the damage is permanent and not always immediately apparent.

SELECTING THE MOST APPROPRIATE CLEANING METHOD

All forms of cleaning can have an adverse effect on the surfaces being cleaned. It is the responsibility of all involved to minimise the extent of damage and risks. Some methods will pose greater threats than others, and the skills of the operatives undertaking the works are of paramount importance. They must also have the correct training, experience, and supervision.

The most appropriate system of cleaning will often involve a combination of methods to deal with the variable nature and extent of soiling on any particular façade.

There are many different methods available for consideration. The selection of the most appropriate methods should initially be made on the basis of those which will minimise the risk of damage. Cost should be a secondary consideration. The risk of damage and costs have to be balanced by the client's requirements and expectations which can often vary from the consultant's and conservation officer's views.

FAR LEFT
The heavily polluted brick façades of Battersea Power Station before the commencement of cleaning, adaptation and repair

LEFT
Corroding steel, brickwork and faience awaiting conservation at Battersea Power Station

ABOVE
Sulphate encrustation at Battersea Power Station

CLEANING

It is generally accepted that it is preferable to 'underclean' rather than risk damaging the surfaces. However, the client's expectations need to be managed due to the fact that they generally expect and require that their building will appear as new. The best way to achieve this is to undertake a range of samples.

Samples should be undertaken in discrete locations across a range of soiled areas to realistically test the effectiveness of the proposed methods and to ensure that the desired results can be achieved without damaging the masonry surfaces.

It is important to understand the façade material being cleaned: limestones react very differently to atmospheric pollutants and cleaning methods than sandstones do. Granites and marbles also will respond differently to alternative methods. Various types of bricks will respond differently to similar cleaning methods.

It is also important to understand the nature of the soiling. This could be atmospheric surface pollutants such as hydro-carbons from diesel exhausts, sulphates from burning coal, chemically bound contaminants such as metal stains, efflorescence, staining from the redistribution of soluble organic residues, lichens and algae, the application of inappropriate treatments, and smoke staining from tobacco and candles. Dissimilar cleaning methods are often necessary to remove different types of soiling.

Alternative methods are often equally viable and it is usually possible to clean any particular façade with a range of methods. Each method, however, will produce different results over various types of

Controlled heated water cleaning of Carrara marble, Leicester Square

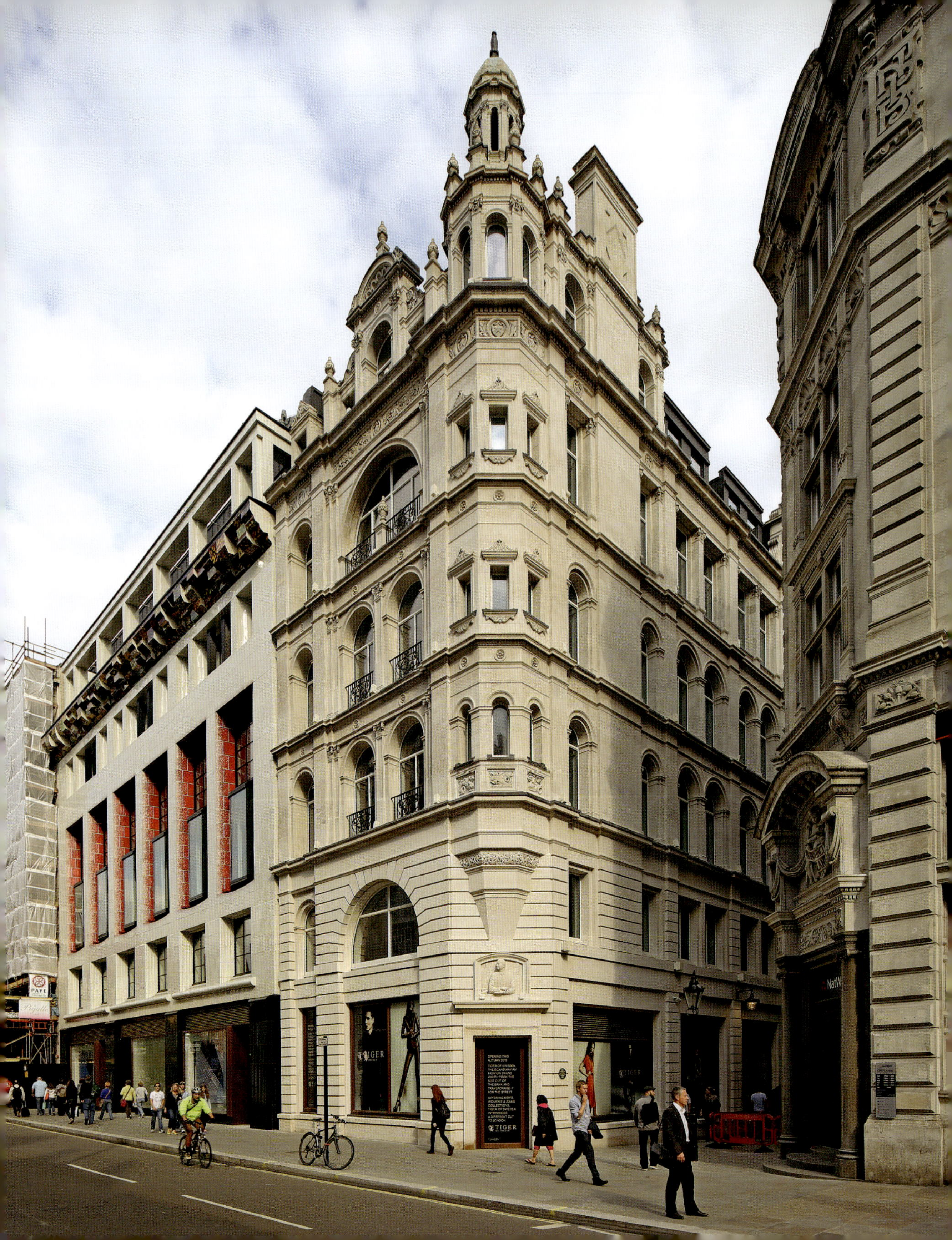
TIGER

CLEANING

soiling which is not usually consistent over a façade or from one façade to another due to the long term effects of pollution and the prevailing wind and rain. In sheltered locations, such as under cills and cornices, soiling accumulates and the resultant concentrations often require different cleaning techniques from those which would be effective over general areas of façade.

Opinions often vary from one consultant to another as to where and when to use a particular method. For instance, many conservation professionals will not consider the use of chemical cleaning agents under any circumstances whilst others routinely specify their use.

The following guidelines are used by PAYE to provide the basis for our recommendations. It is likely that other opinions might be equally valid and results might vary in consequence. Our proposals have never been rejected by local conservation officers or English Heritage, always provide value, and on the whole meet client's expectations.

THE VARIOUS METHODS

The various methods ranked on a cost per m² basis starting with the cheapest are set out below. Note that not all of these would be recommended by PAYE for use on historic masonry. These methods are explained in detail below.

- High pressure water lance (generally not recommended)
- Fine nebulous water sprays
- Chemical cleaning (generally not recommended)
- Pressurised hot water/steam
- Grit blasting (not recommended)
- Low pressure air abrasion
- Hand sprayed nebulous water and swabs/sponges
- Poulticing

HIGH PRESSURE WATER LANCE

This method is generally not regarded as appropriate due to the risk that the high pressures will physically damage the masonry surfaces, and that potentially harmful deposits could be driven further into the pore structures. If unsupervised, unscrupulous companies sometimes use this method having previously agreed costs on the basis that a more appropriate and costly technique will be used.

LEFT
Cleaned, dismantled and reconstructed Portland stone façades in Piccadilly featuring new lower two storeys designed by Eric Parry Architects

ABOVE
Cleaned and repaired mosaic and marble details, Regent House, 235 Regent Street

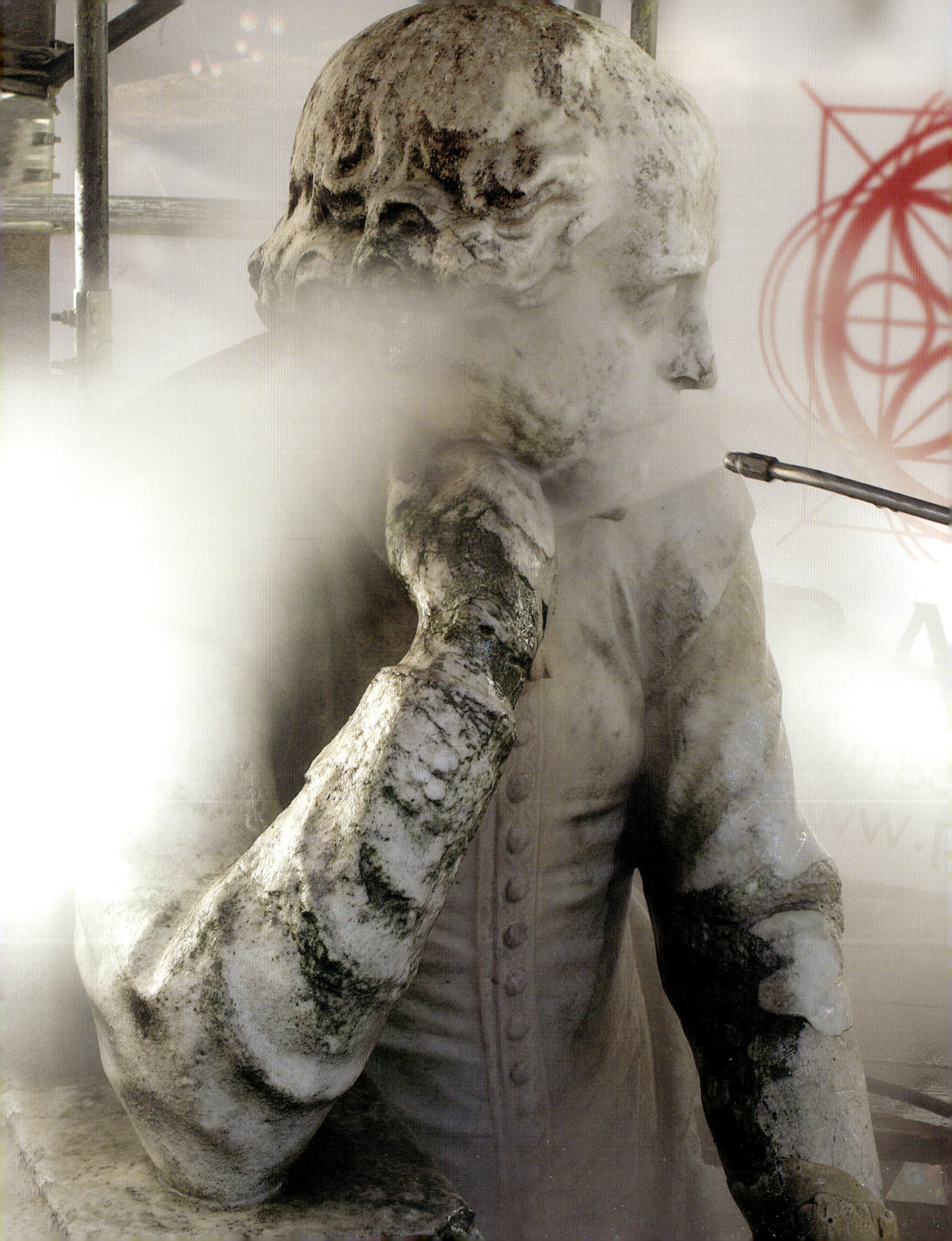

CLEANING

FINE NEBULOUS WATER SPRAYS

This method is sometimes also referred to as the traditional masons' water cleaning method. It involves the use of mains pressure water with soft bristle brushing and is particularly effective on limestones due to the fact that these stones are water soluble and therefore dissolve slightly when wetted enabling the contaminants to be released from the surface. This method will not work on sandstones and other materials which are not water soluble.

The volume of water must be managed carefully to avoid unnecessary saturation and inconvenience to pedestrians and users of the building. Typically, temporary gutters will need to be constructed to collect the water and manage its disposal. Filters will be necessary to ensure debris does not enter and block the drains.

The risk of water ingress is high so precautions must be taken around window and door openings by installing heavy duty polythene secured with low tack adhesive tape. An inspection of other potential defects which would enable water to leak into the building must also be taken before the water is turned on, and further precautions taken such as the repointing of open masonry joints.

Occasionally we are asked to use timers to control the periods of wetting. This is a throwback to the 1980's when this was considered a potential means to control the amount of water so that over-saturation could be avoided. In practice, this will not control saturation but actually leads to the masonry absorbing greater volumes of water and extends the amount of time taken to achieve the desired result. At the Palace of Westminster, PAYE were requested to limit the use of water to one hour in every three in any particular area of façade (an hour was sufficient to clean the masonry).

In another example of an inappropriate specification, PAYE were once requested to clean a limestone façade using de-ionised water. We were told by the architect that 'it seemed like a good idea' but this requirement was dropped after it was realised that de-ionised water actually risked accelerating the corrosion of the embedded cramps.

It is not unusual for a brown stain to develop on limestones during nebulous water cleaning. This is caused by the migration of soluble naturally-occurring organic residues to the masonry surface as the saturating moisture evaporates. This problem tends to occur in areas where the residues have previously been mobilised

LEFT
Carrara marble statue of Shakespeare being carefully cleaned by means of steam, Leicester Square

ABOVE
Portland stone statue of Shakespeare's Prospero by Eric Gill being cleaned with water and soft bristle brushes, Broadcasting House

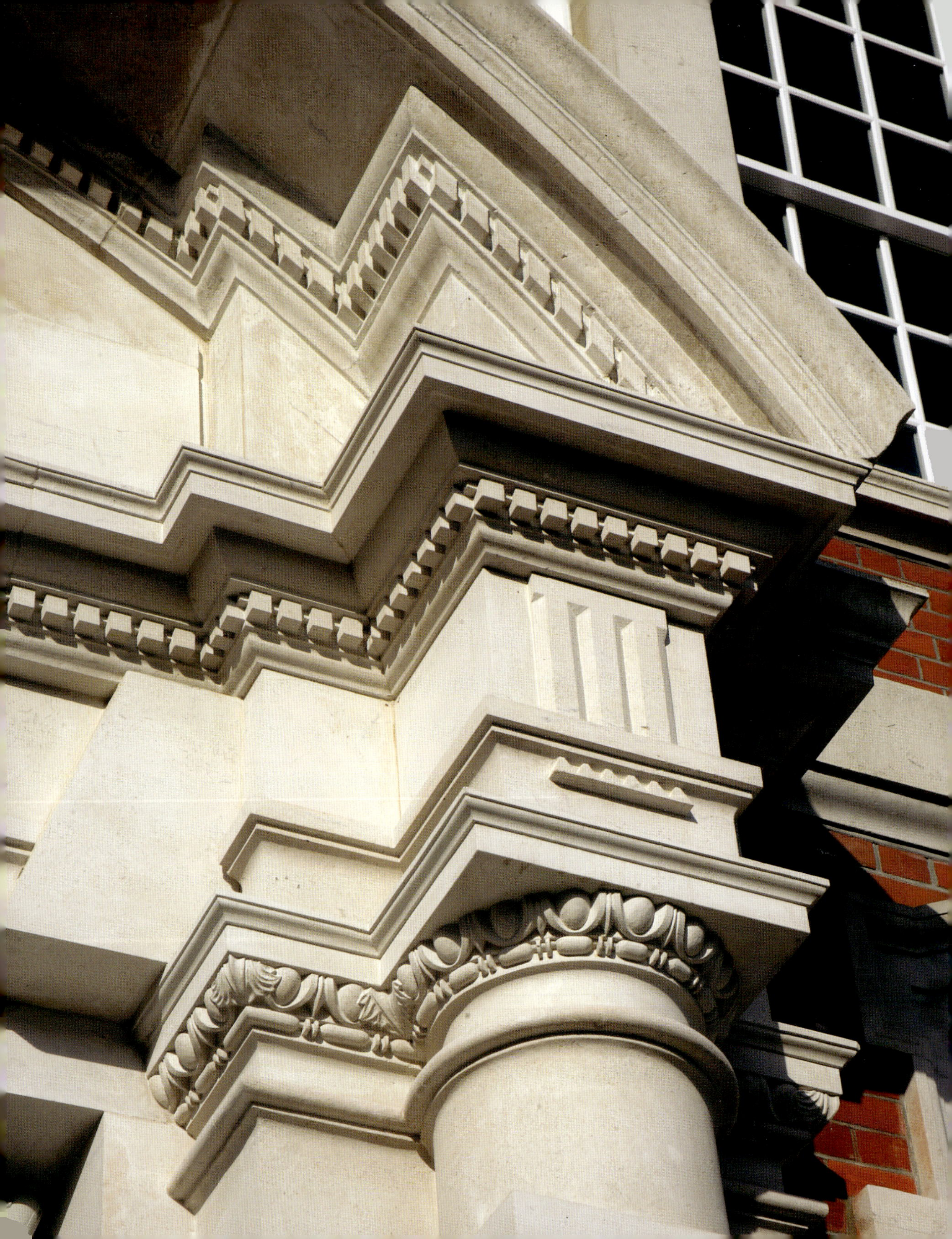

CLEANING

by water from a leaking rainwater pipe or gutter. The stain will break down under long term exposure to UV light. If we are required to remove the stain as part of the cleaning process, poulticing techniques will need to be used (see below).

It is occasionally appropriate to minimise the volume of water by using hand sprays with sponges or swabs on particularly sensitive or delicately carved limestone.

CHEMICAL CLEANING

Methods involve either the use of an acid (typically hydrofluoric acid) or an alkaline (typically sodium hydroxide). Acids will adversely react with limestones and also have the potential to damage brickwork and some sandstones. Alkalines can cause the formation of salts on stone or brick surfaces. In our opinion, the use of either acid or alkaline cleaning chemicals on natural stone is best avoided.

Acids can be effective in cleaning terra cotta, faience, and some bricks but extreme care is needed to avoid the risk of damage.

Alkalines can be used as a degreasing agent to clean faience and granites. Some cleaning systems use alkalines to clean surfaces with acid neutralising agents as a subsequent after-treatment.

Acids have the potential to etch glass, and alkalines will damage paintwork. Polythene and tape protection is therefore essential when using chemical cleaning methods.

Today hydrofluoric acid is commercially supplied at 12% solution and at PAYE we find that this can be further diluted down to less than 2% and still be effective. In the 1970's and 1980's however, hydrofluoric acid was taken to site at 70% solution for dilution at the discretion of the cleaning operatives. At these strengths hydrofluoric acid is extremely dangerous, and its use on scaffolding over public pavements could not be considered now due to safety concerns. Many terra cotta and faience buildings were damaged during this period as is evidenced by the vertically streaky appearance of acid burns on these façades.

LEFT
Cleaned and indent repaired Portland stone portico, Chelsea

ABOVE LEFT
Heavily sulphate encrustation on Anston stone and corroding ferramenta at the Palace of Westminster

ABOVE
Carefully controlled fine nebulous water cleaning of Portland stone, Whitehall

CLEANING

ABOVE
Cleaning of marble at Fitzroy Place Chapel

RIGNT
Cleaned, reconstructed, and repaired faience façades at the Regents Palace Hotel, Regent Street

When using chemicals on absorbent surfaces it is first necessary to pre-wet the surfaces to control the absorption and keep the active chemical on the surface. Dwell times should be minimised prior to washing off with copious amounts of clean water.

The chemistry of natural stones varies considerably from one type to another. Some stones will react strongly with some chemicals. Unfortunately, manufacturers of chemical cleaning treatments do not adequately differentiate between various stone types, so the potential for damage is therefore high in the hands of inexperienced and untrained operatives. In Scotland during the 1980's, there was a complete moratorium on façade cleaning following the rapid degradation of many sandstone façades after they had been inappropriately cleaned with strong acids.

At PAYE we would generally advise against the use of chemicals for cleaning limestones and sandstones. The chemical cleaning of granites, brickwork and terra cotta can sometimes be appropriate if used with care and consideration.

Chemicals should always be used at the minimum possible solution strengths and washed off after a minimal dwell time. Samples should always be undertaken prior to commencement in order to establish the minimal solution strengths and minimal dwell times.

Suitable PPE is essential and typically will include appropriate goggles, gloves/gauntlets, rubber steel-capped boots, and waterproof overwear.

PRESSURISED HOT WATER/STEAM

Water pressures and temperatures can be controlled by the operative between hot water and steam to achieve the best balance for dealing with varying situations on the façade. Temperatures can be increased by reducing the pressure of the water.

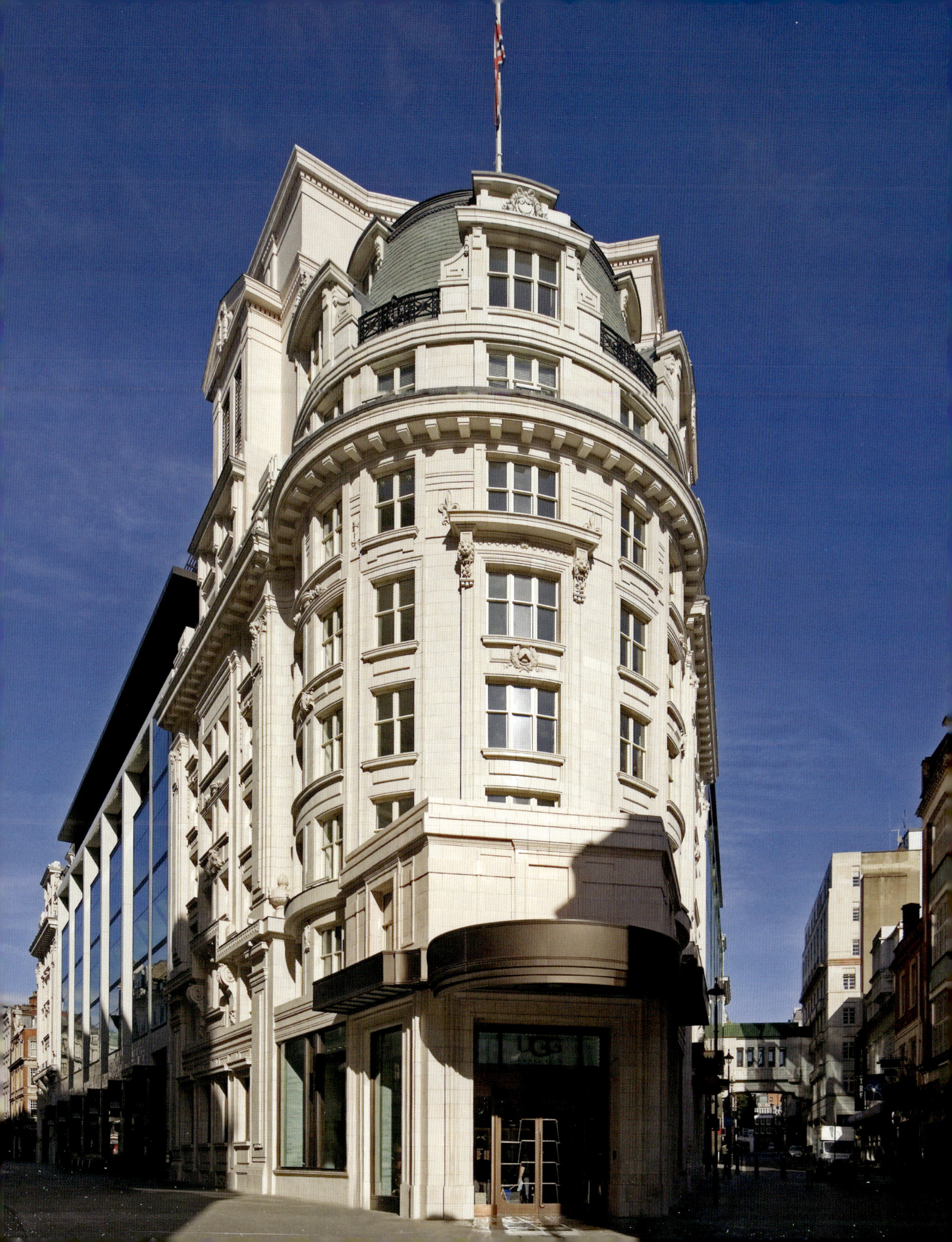

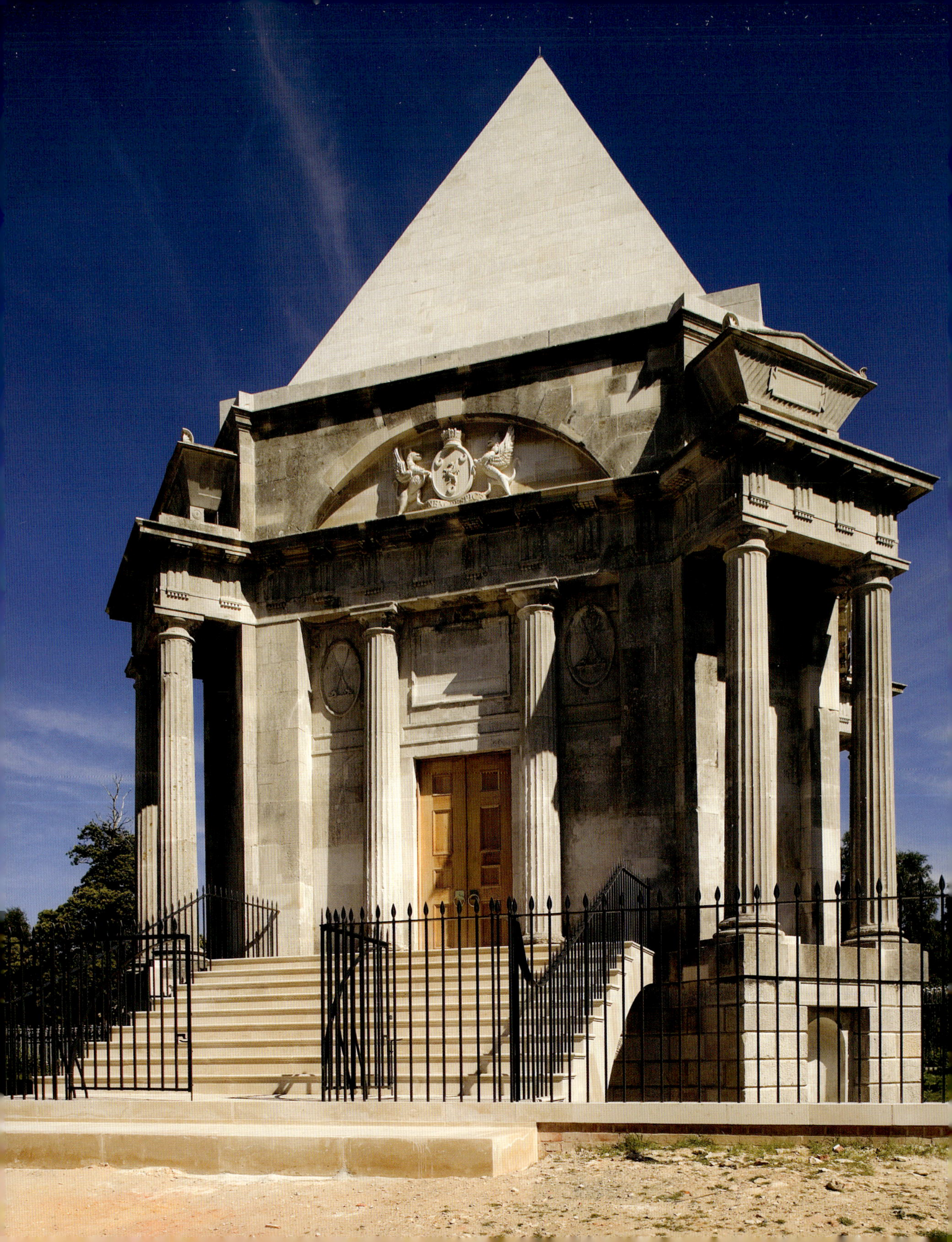

CLEANING

This method is a useful alternative to fine nebulous water sprays where the higher water temperature helps to remove surface contaminants but the high volumes of water are avoided. The level of cleaning tends to be not as effective as that achieved using fine nebulous water sprays. This can be considered an advantage on certain conservation projects where a 'deep' clean is not required.

The use of hot water/steam cleaning can be effective on brickwork, particularly on dense engineering bricks.

This method is also effective to supplement other methods such as pre-cleaning prior to the application of chemicals, and for later washing off.

We have also found that hot water/steam can be an effective treatment for removing algae (hot water acts as a biocide). Such treatments need to be repeated periodically.

GRIT BLASTING

Grit blasting involves the use of a grit propelled towards the surface by means of compressed air. The larger the particle size, the harder the grit, and the greater the pressure (ie particle speed), the greater will be the risks that the masonry surfaces will be damaged. Sands are no longer used as abrasives due to the dangerous nature of the dust created (silica dust causes silicosis). Graded aluminium oxides are often used for grit blasting today.

The potential for damage is obviously high. Surfaces will be cleaned quicker and costs will be reduced by maximising the pressures and using coarser and harder grits. The reality is that the lower the cost, the greater will be the risk of damage.

LOW PRESSURE AIR ABRASION

This is a more sophisticated form of grit blasting using softer and finer abrasives and controlled minimised pressures. A typical abrasive used for this process is a very finely graded calcite dust. Pressures can be effective as low as 2 bar or less.

Risks are minimised by using the finest abrasive particle, using a soft abrasive, minimising the pressures involved, and controlling the minimum distance between the nozzle and the masonry surface.

It is good practice to also use water to control the amount of dust generated and to help cushion the abrasive. Dry grit blasting is best avoided unless there are specific reasons for water not to be used.

LEFT
The Portland stone façades of the Darnley Mausoleum were not cleaned as part of the conservation works

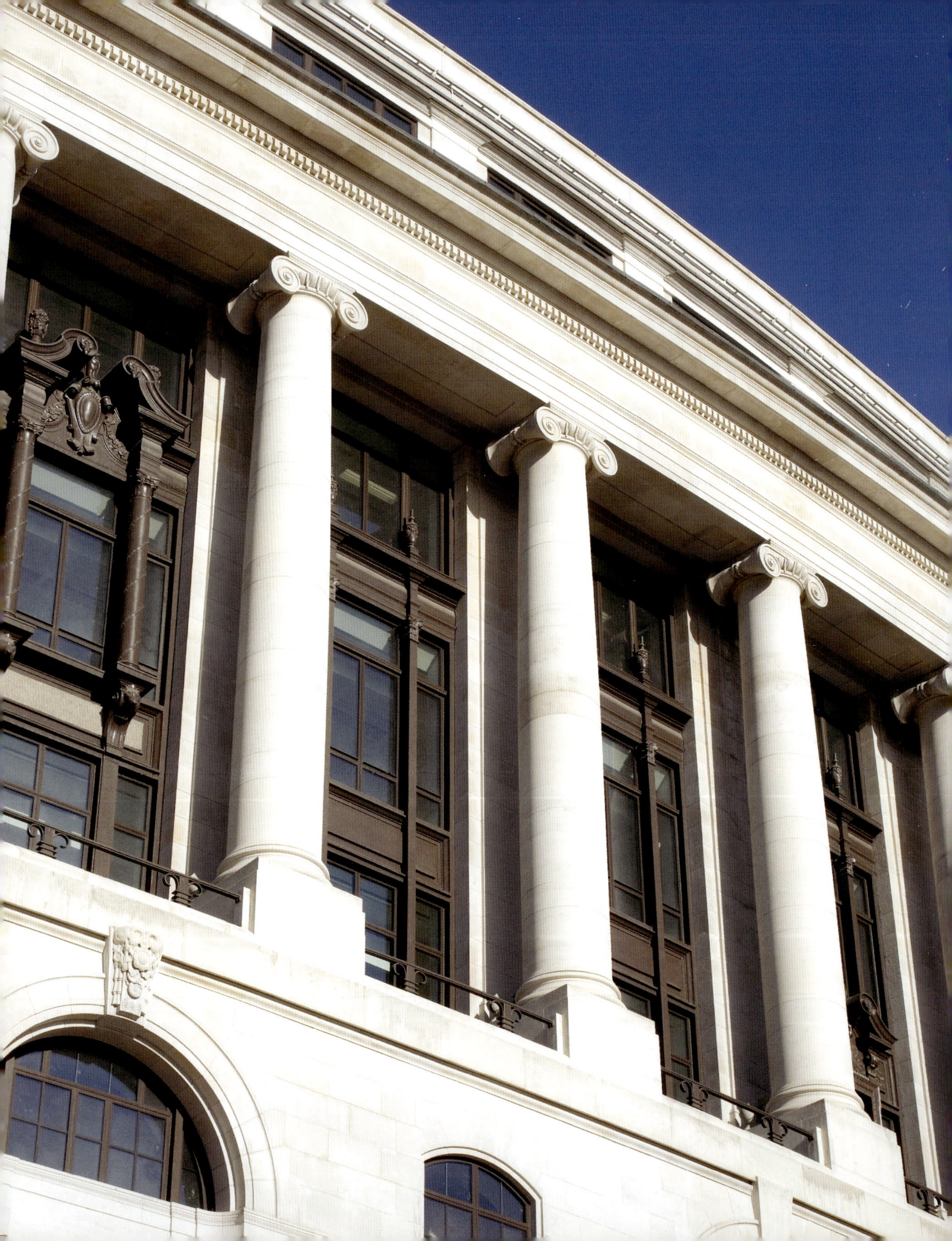

CLEANING

ABOVE
Cleaned and repaired Bath stone at Clapton Portico, Hackney

LEFT
Cleaned and repaired Portland stone façades, Unilever House

CLEANING

ABOVE AND RIGHT
Cleaning of Portland stone using latex poultice at Fitzroy Place Chapel

FAR RIGHT
Cleaned and repaired Portland stone, Whitehall

POULTICING

Poulticing is the application of a paste, sometimes containing formulated chemicals. They draw contaminants from the masonry pore structure into the paste which is subsequently removed. Many poultices are based on cellulose, but others use lime putty, paper pulp or clay. Surface dwell times can be as low as one hour, but could be as long as many days. Some poultices must not be allowed to dry so it is sometimes necessary to cover them with thin polythene sheet or clingfilm. Poultices are typically used to reduce metal stains or remove unsightly soluble organic residues from limestone surfaces following saturation from leaking rainwater pipes.

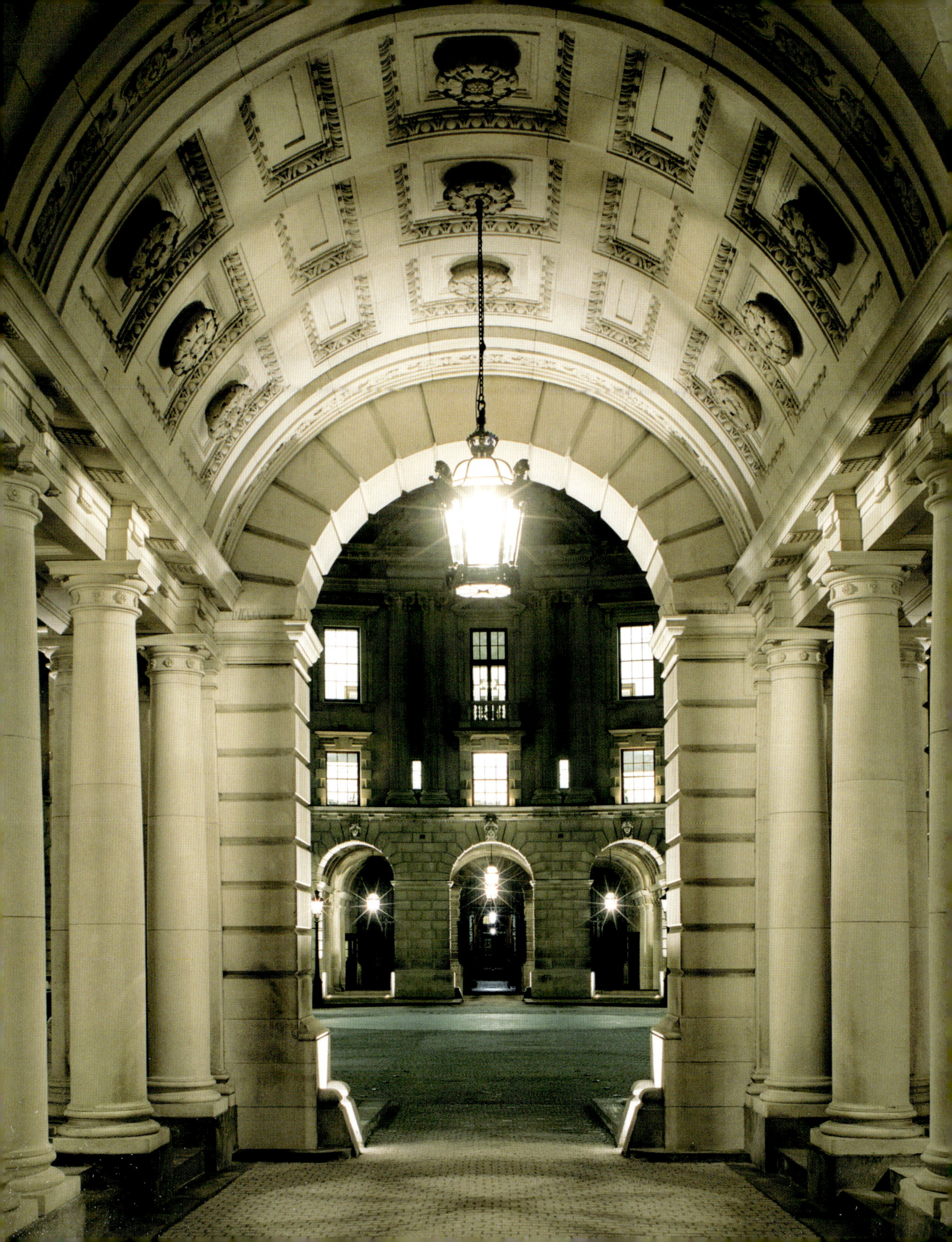

TEN TRINITY SQUARE

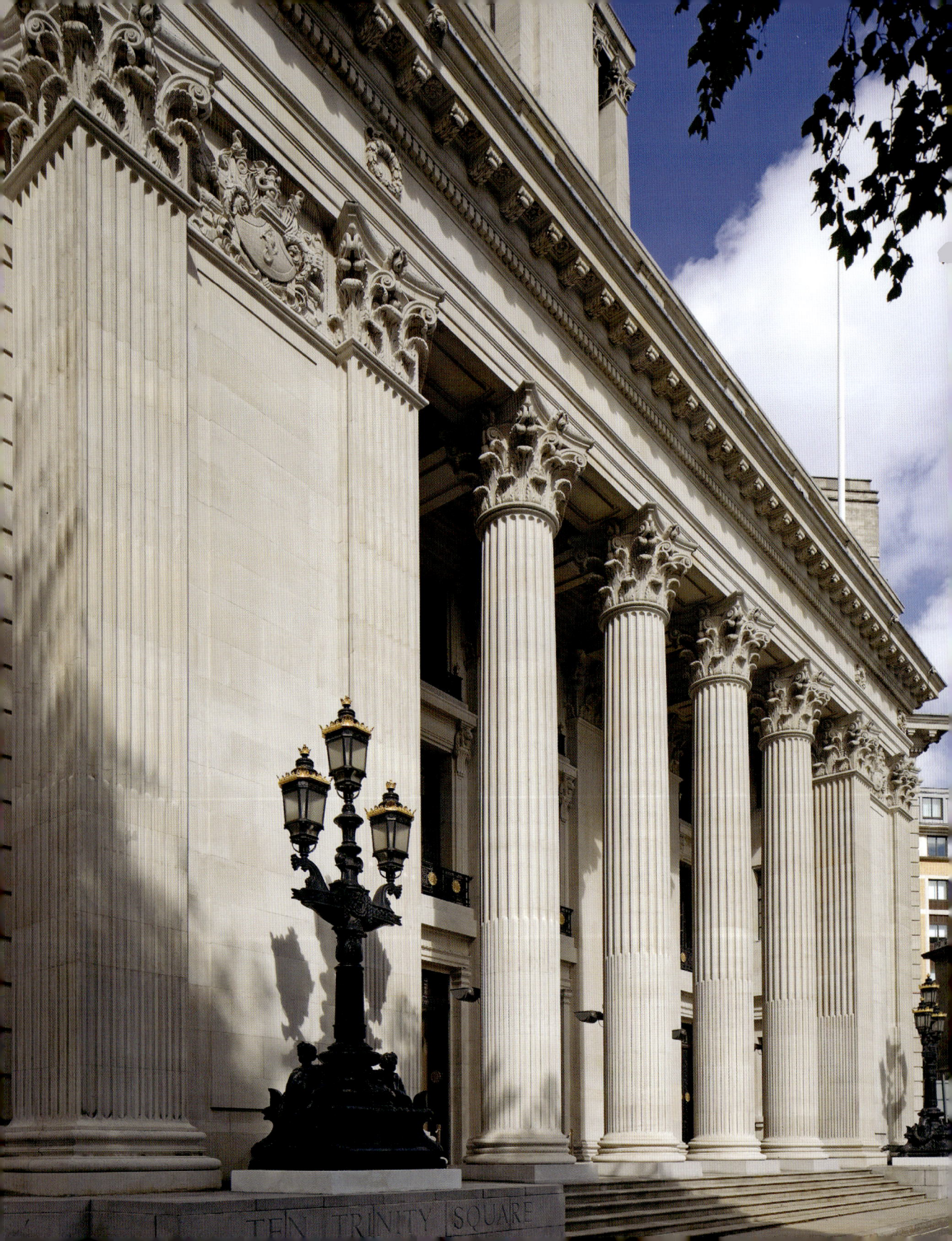

CLEANING

METHOD REFERENCE GUIDE

The following is a reference guide for methods which can generally be considered for use upon the various surfaces materials:

LIMESTONES - fine nebulous water sprays, hot water/steam, low pressure air abrasion, and poulticing. Avoid the use of chemicals.

SANDSTONES - hot water/steam, low pressure air abrasion, and poulticing. Avoid the use of chemicals.

MARBLES - fine nebulous water sprays, hot water/steam, poulticing. Avoid the use of chemicals.

GRANITES - hot water/steam, chemical cleaning agents, low pressure air abrasion, and poulticing.

RECONSTITUTED STONE - fine nebulous water sprays, hot water/steam, low pressure air abrasion.

BRICK AND TERRA COTTA - hot water/steam, low pressure air abrasion, controlled use of acid chemicals.

GLAZED BRICK AND FAIENCE - controlled use of alkaline chemicals (sometimes with acid after-washes if deemed necessary to neutralise any residues), hot water/steam.

SEQUENCING

Cleaning is an operation which needs to be carefully managed to avoid being disruptive to other trades and building users. In our experience it is always better to complete the cleaning work as early as possible into the sequence of operations, but not necessarily during excavation work adjacent to the façade particularly where larger volumes of water are involved. Cleaning at an early stage helps in the selection of appropriate mortars and matching suitable stones and bricks.

It is always preferable to clean a façade in one operation because break lines between phases can be visible. If it is necessary to undertake the work in phases, it is important to clean each phase to a suitable break line such as a return, corner, or if nothing else is available, a rainwater pipe.

ABOVE
Cleaned and repaired Portland stone at Edwin Lutyens' Portland stone façades of 85 Fleet Street

LEFT
Cleaned and repaired Portland stone in Trinity Square

59

CLEANING

Above and Right — The glazed brick and Portland stone façades of one of the many courtyards within the HM Treasury, Whitehall: before and after cleaning, adaptation and conservation.

FINAL CLEAN

Concerns are often expressed that if the façade is cleaned at the commencement of the repair programme it will become soiled during the subsequent repair and conservation works, so good practice dictates that a final light clean should be undertaken immediately prior to the scaffold being struck to remove any dust that has settled on the façade during the works.

An effective method to remove dust as a final clean would be a low to medium pressure water lance, always maintaining a minimum permitted nozzle distance from the masonry to avoid the risk of damage.

'Board Rash' is a term used to describe the areas of façade that will become soiled by rain splashing off dusty scaffold boards. This can be removed during the final light clean prior to the scaffold being struck.

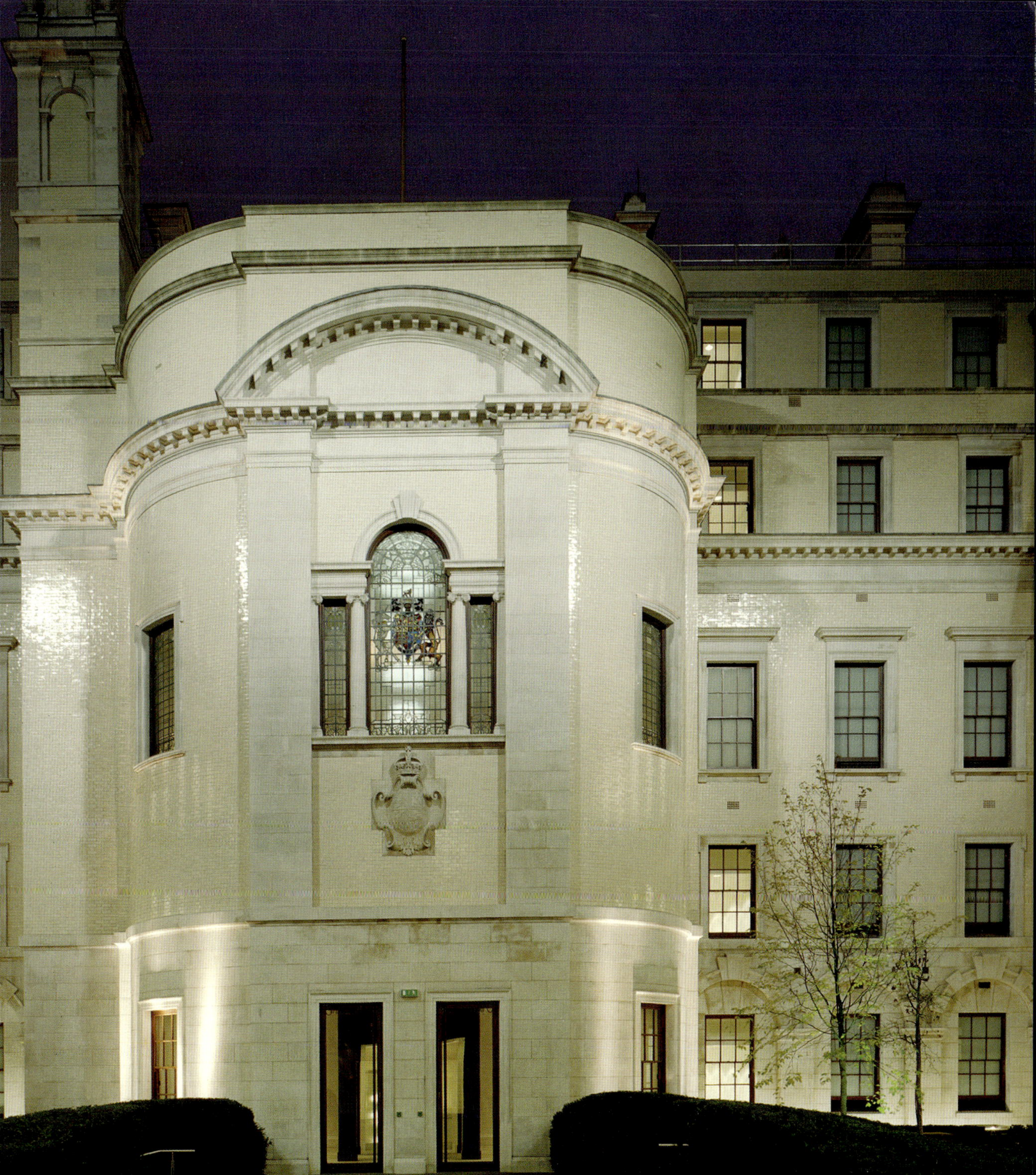

2005 - 2007
Restoration

William Benyon
Englefield Trust
Wroughton
Lieutenant

2006

2005 - 2007
Restoration

Lady Elizabeth Godsell

John Madejski

CHAPTER FOUR

REPAIRS

Developing a scope of conservation and repair works must start with a thorough investigation and understanding of the causes of failure (see Chapter 2). Once these have been identified and measures taken to nullify or at least minimise their impact on the masonry, the detailed scope of works can be developed. At this stage it is important to understand the client's requirements and objectives. Lower cost expenditure tends to equate to short term maintenance periods, whilst spending more on repairs will generally provide for longer maintenance periods. Cost analyses suggest that the most economic strategy is usually the long term view, which obviously suits clients with long term interests in their properties. The ability to pass on the maintenance responsibility to tenants will sometimes encourage certain clients to spend less on repairs than perhaps they should.

The first step is to understand the client's requirements for planned maintenance periods.

Bath stone and flint repairs at Reading Minster

PLANNED MAINTENANCE PERIODS

It is generally appropriate to plan future façade cleaning and repair works to coincide with other future repairs eg. window redecoration, mastic sealant replacement works, etc. This will optimise the future use of scaffolding and thus minimise costs. If it is planned that a façade will be scaffolded every 10 years, then, unless it is required or appropriate otherwise, specific repairs need only be undertaken where the stone will not safely perform over this period.

The replacement of stone on the façade of an occupied building will cause a degree of inconvenience so it is usually considered appropriate to undertake this work before it is occupied if possible. For this reason, extended maintenance periods in excess of 20 years are often required after a major façade refurbishment project. Obviously, it will be necessary for inspections to be undertaken at frequent intervals during such extended periods. Any necessary minor maintenance work, such as the repointing of masonry joints on coping stones, must be undertaken within these periods to ensure that defects do not develop and become dangerous. It could not be considered responsible to assume that an historic masonry façade will perform for an extended period of two or more decades without regular inspections and minor maintenance.

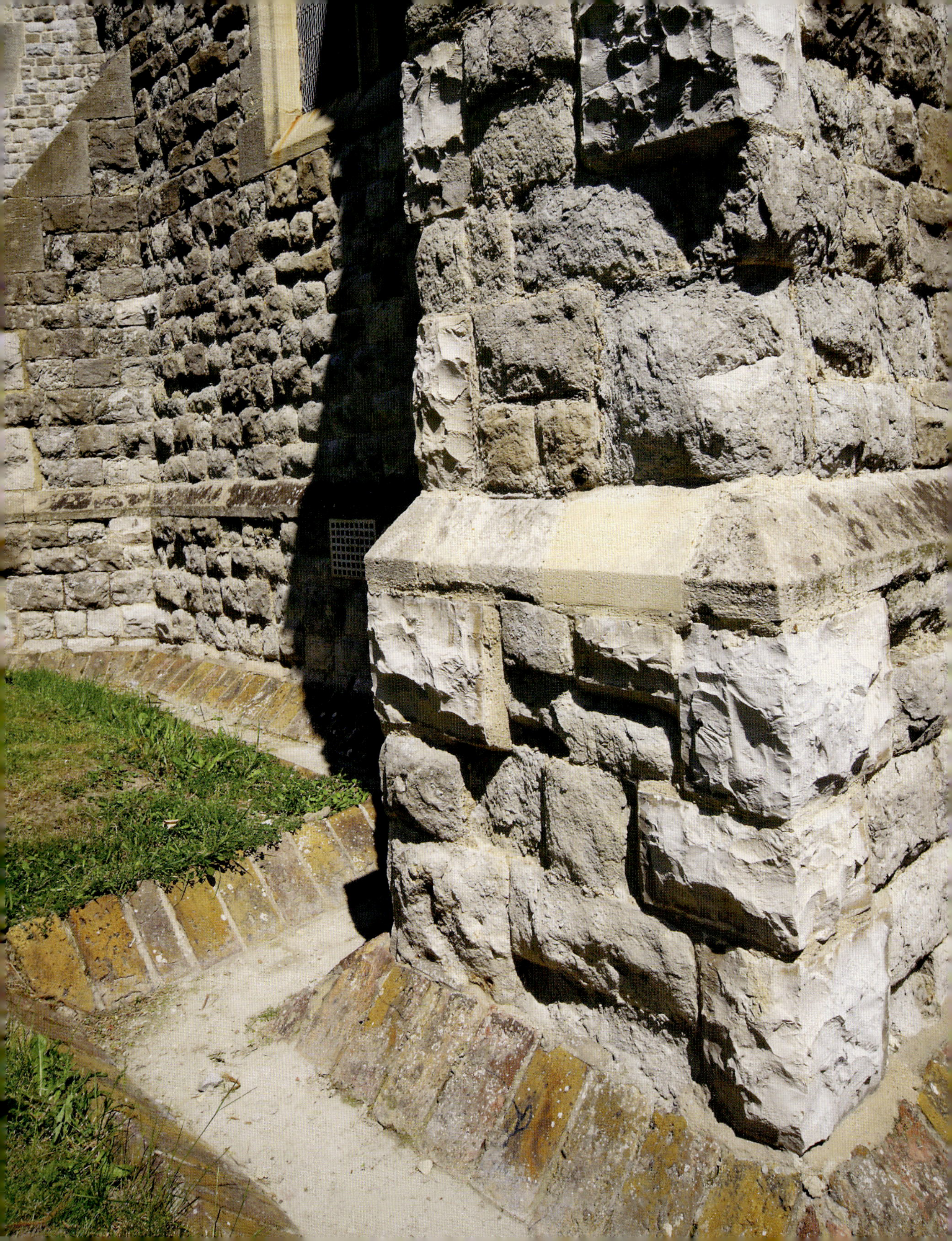

REPAIRS

Operating and maintenance manuals are usually required upon completion of a project and PAYE will provide recommendations for inspection routines and planned maintenance works, if requested.

When developing a proposed scope of works, it is important to understand the client's requirements for costs to be minimised in either the short term or the long term. Some defects, such as the gradual loss of profile from the effects of weathering, manifest over long periods of time and can often be left until subsequent phases of maintenance and repair without risk. Others, such as displacement and fracturing caused by the corrosion of embedded metalwork, may develop quickly particularly if exacerbated by a blocked or broken rainwater pipe. Such a situation can soon develop to the point where safety

FAR LEFT
Kentish Ragstone and Bath stone indent repairs and repointing at St Saviours, Westgate, Kent

ABOVE
The Bath stone and flint façades of Reading Minster

LEFT
Operating the hydraulic ram to slide forward the façade of Chenil House, Chelsea

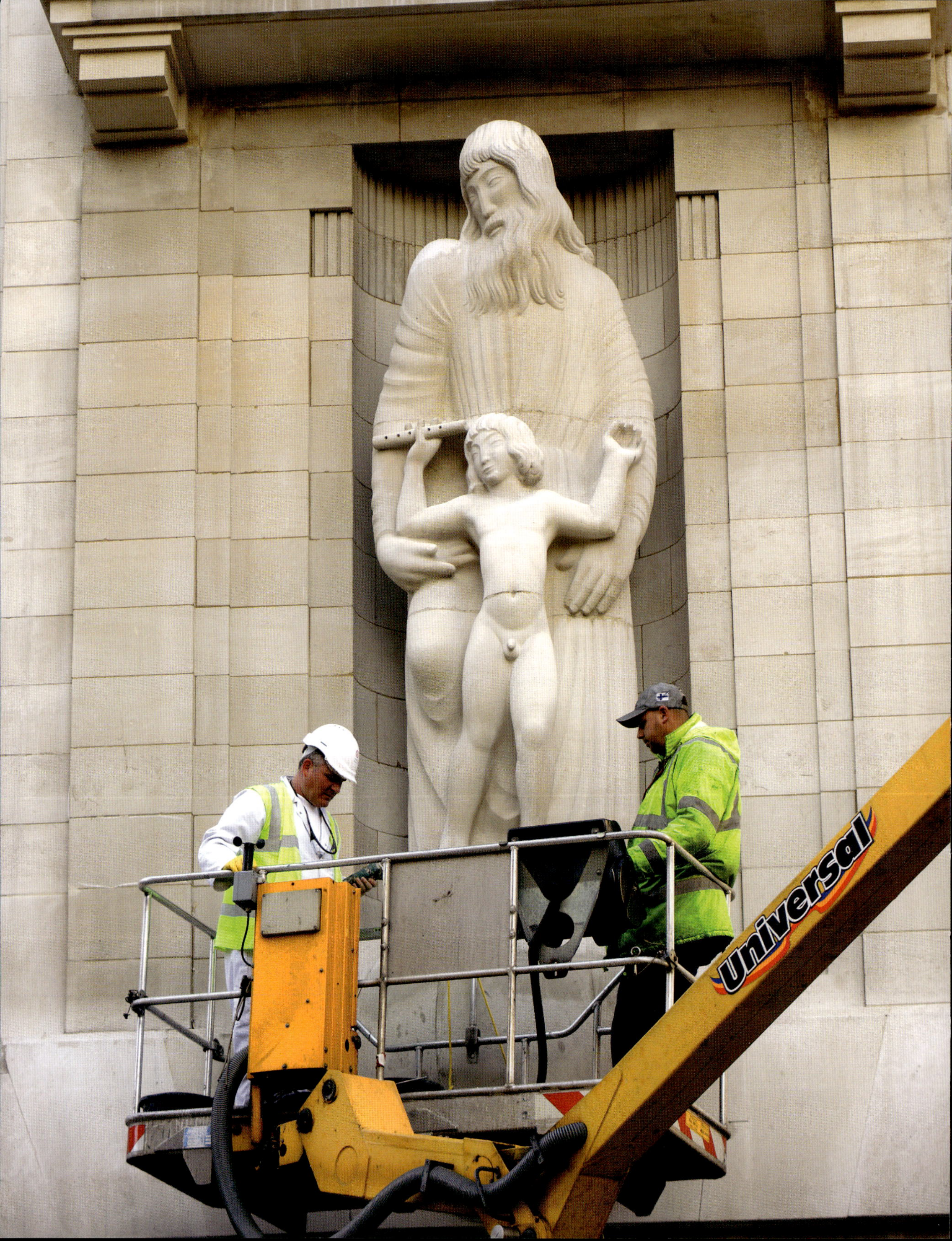
Universal

is compromised and therefore needs to be properly identified and rectified as a matter of urgency.

Saving costs in the short term will usually result in the need for additional repairs at a later date, particularly if the cause of failure is not properly addressed. Deferring major work until a future date needs to be carefully considered, especially if future repairs will be disruptive to building users and involve multiple specialist trades. Safety is also an important consideration. For this reason, it is often more appropriate to carry out a full scope of work during major refurbishment, particularly if the building is not occupied at the time.

SURVEYS AND IDENTIFYING THE CAUSES OF FAILURE

It is important that detailed and accurate surveys are undertaken at the earliest opportunity within the overall works programme. Usually this is best achieved from scaffolding, but it is often possible to produce reasonably detailed and accurate surveys from aerial platforms or by using binoculars from pavement level. Inspections from roofs and windows are also helpful. Sometimes it is appropriate to employ qualified abseilers to identify, schedule and photograph defects, particularly if trees obstruct binocular surveys or prevent aerial platforms from getting sufficiently close to the façades.

An accurate survey should identify all defects and if possible also show where previous defects have been rectified. This will help to identify locations on the façade which are susceptible to failure, which in turn will help to identify the causes.

In developing a proposed scope of work, the most important issue is to identify and understand the causes of failure. Value cannot be provided by merely undertaking repairs unless the causes of the defects are also identified and rectified. For instance, at BBC Broadcasting House in London, PAYE identified a poor parapet detail as the cause of corrosion of the embedded steel frame in many locations. Major expensive repairs in these areas had been undertaken frequently throughout the building's 70 year life, but none of them had actually solved the problem which was simply the inadequate protection of embedded steel parapet beams against rainwater ingress. Working with the project architects, PAYE were able to develop a new parapet detail for which listed building consent was granted. The enhanced detail was sympathetic to the original design and not visible from ground level. It has dramatically reduced future maintenance costs and risk liabilities.

The causes of failure often involve a general lack of maintenance, inadequate rainwater management, inadequate

LEFT
Inspecting Eric Gill's Portland stone statue of Prospero and Ariel at Broadcasting House during conservation work

ABOVE
Surveying the Eleanor Cross, Charing Cross Station

REPAIRS

maintenance of the rainwater disposal system, the absence of adequate protection over cornices, inappropriate treatments and inappropriate previous repairs. Rectification often involves other specialist skills such as leadwork as well as traditional masonry.

RAINWATER MANAGEMENT

The vast majority of defects are caused by ineffective rainwater management such as the inadequate maintenance of gutters and rainwater pipes or poor detailing. It is therefore essential that all defects and inadequacies in the management of the rainwater disposal system are identified and incorporated into the repair programme. It is also generally considered appropriate to protect stone from repeated saturation by installing protective lead cappings on cornices (and possibly copings too if the stone is of poor durability). Adequate drips must be provided and these can be achieved in a number of ways, some being more discrete than others.

ORDINARY PORTLAND CEMENT (OPC)

Mortars containing OPC are much less permeable than most building stones and lime mortars. The use of OPC in mortars will therefore generally result in moisture being trapped and concentrated around the bedding, pointing or repair, and this in turn will lead to the localised degradation of the masonry. This defect will manifest itself on most types of stones except the more impermeable granites.

Another issue concerning the use of OPC is that these mortars will have the potential to shrink back from the stone surface, potentially resulting in fracturing of the pointing or repair. Rainwater will

RIGHT
The damaged brick façades of the Elizabeth Garrett Anderson Hospital (now Unison House), Euston Road prior to cleaning and conservation

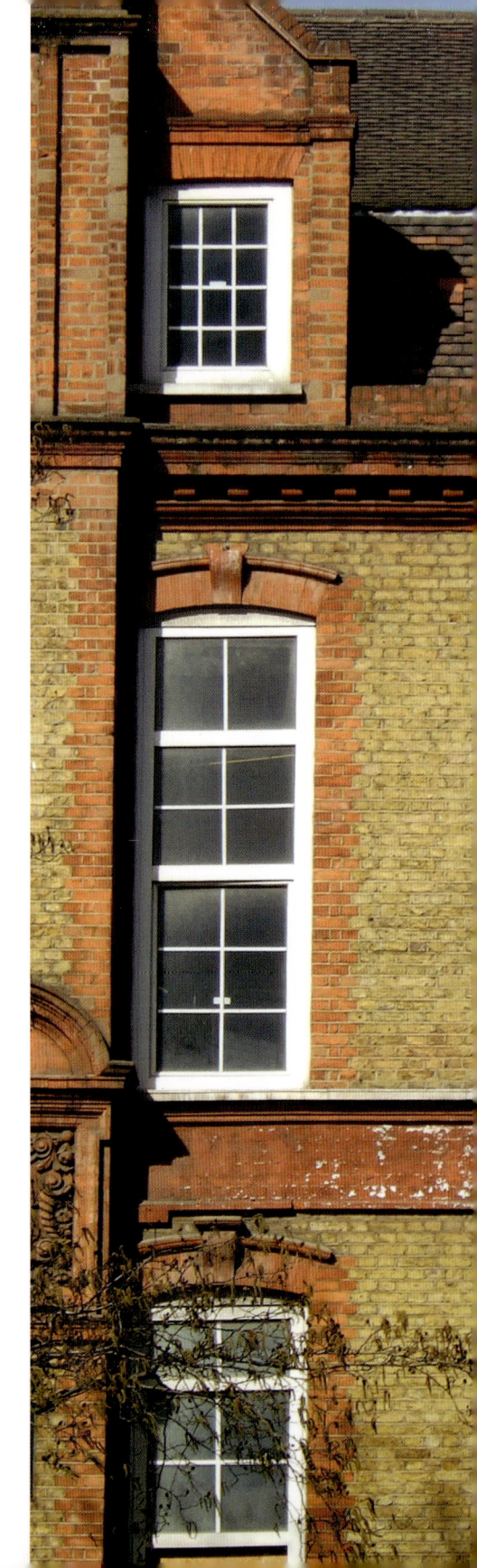

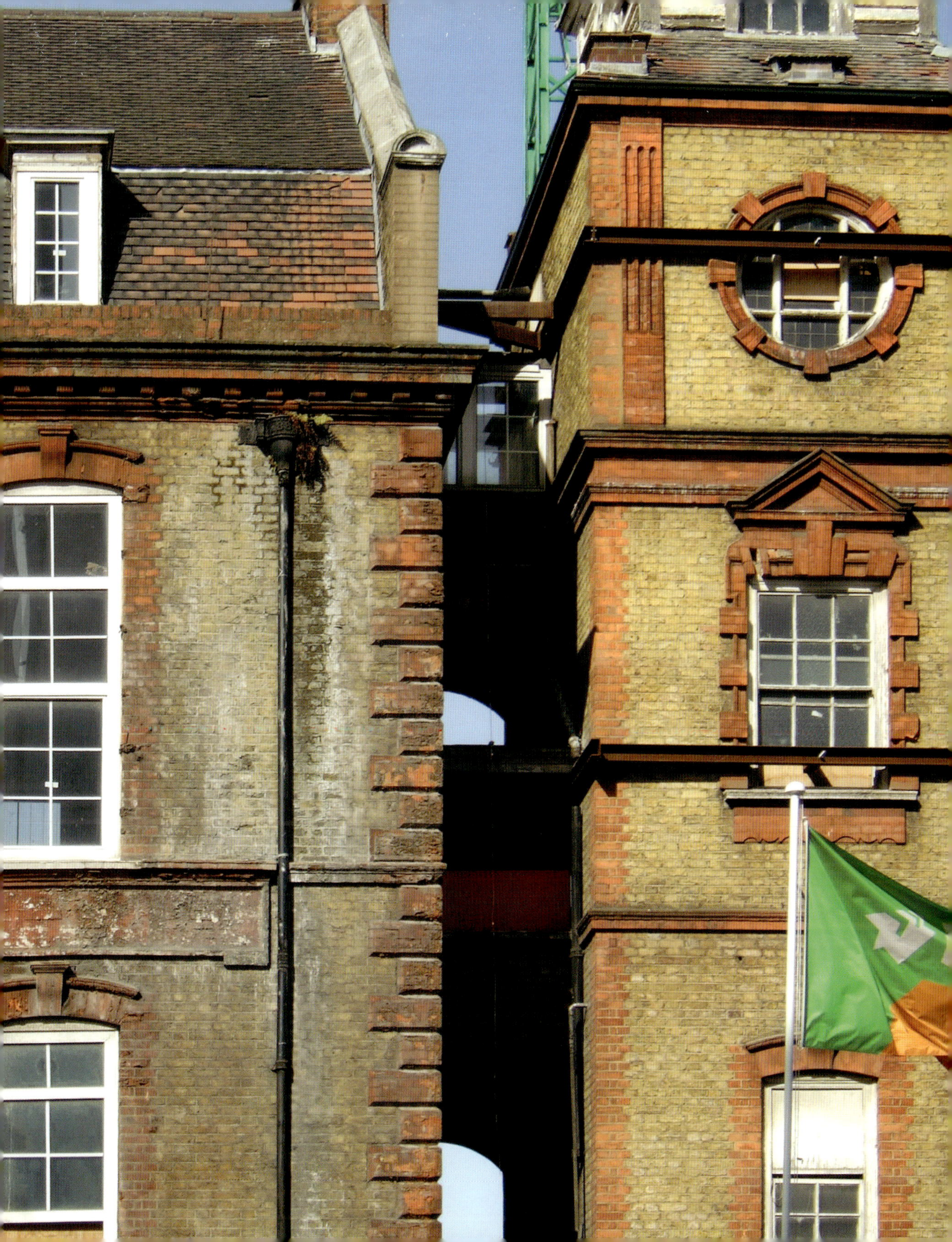

Above
The cleaned and repaired Edwin Lutyens' Portland stone façades of 85 Fleet Street

Right
Portland stone façades of the 18th century Darnley Mausoleum

consequently have the potential to penetrate through these cracks and locally saturate the masonry, leading to further degradation. This effect will be worse if the masonry has been installed with hollow perpend joints (a perennial problem as old as the craft, and found on masonry of all ages from medieval times).

More impermeable materials such as faience and granite will be less susceptible to damage resulting from the use of OPC mortars. In many European countries, granite is often bedded and pointed using OPC mortars with little or no apparent detrimental effect.

REPOINTING

Bedding and pointing mortars should always be more permeable than the stones and bricks they adhere together. This is to ensure that moisture is drawn into the joint so that the mortar degrades in a sacrificial way, thus protecting the masonry. This is preferable because it is always more economical to replace the mortar than the stones or bricks. If the mortar is less permeable than the masonry, the masonry will degrade.

On most historic structures the mortars have been carefully selected to provide an adhering rainwater and wind resistant bonding material which will degrade quicker than the stones and bricks. It should therefore be expected that every maintenance programme will always involve some repointing works, particularly on the more exposed areas of the building such as the cornices, copings, plinths and cills.

In the 1980's most specifications required mortars to be removed to a minimum depth of 15mm as part of the preparation of the joint for repointing. English Heritage Technical Guidelines at

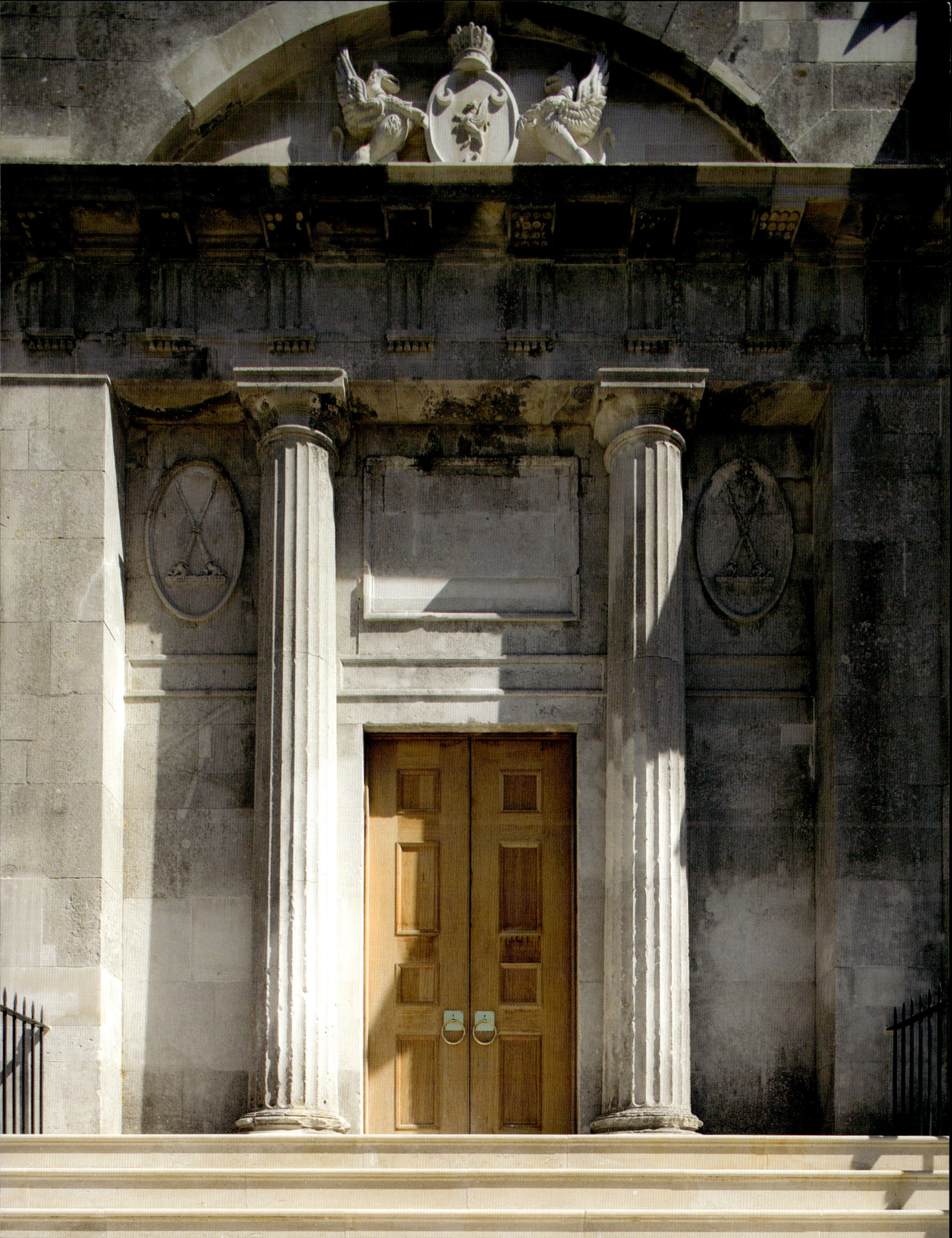

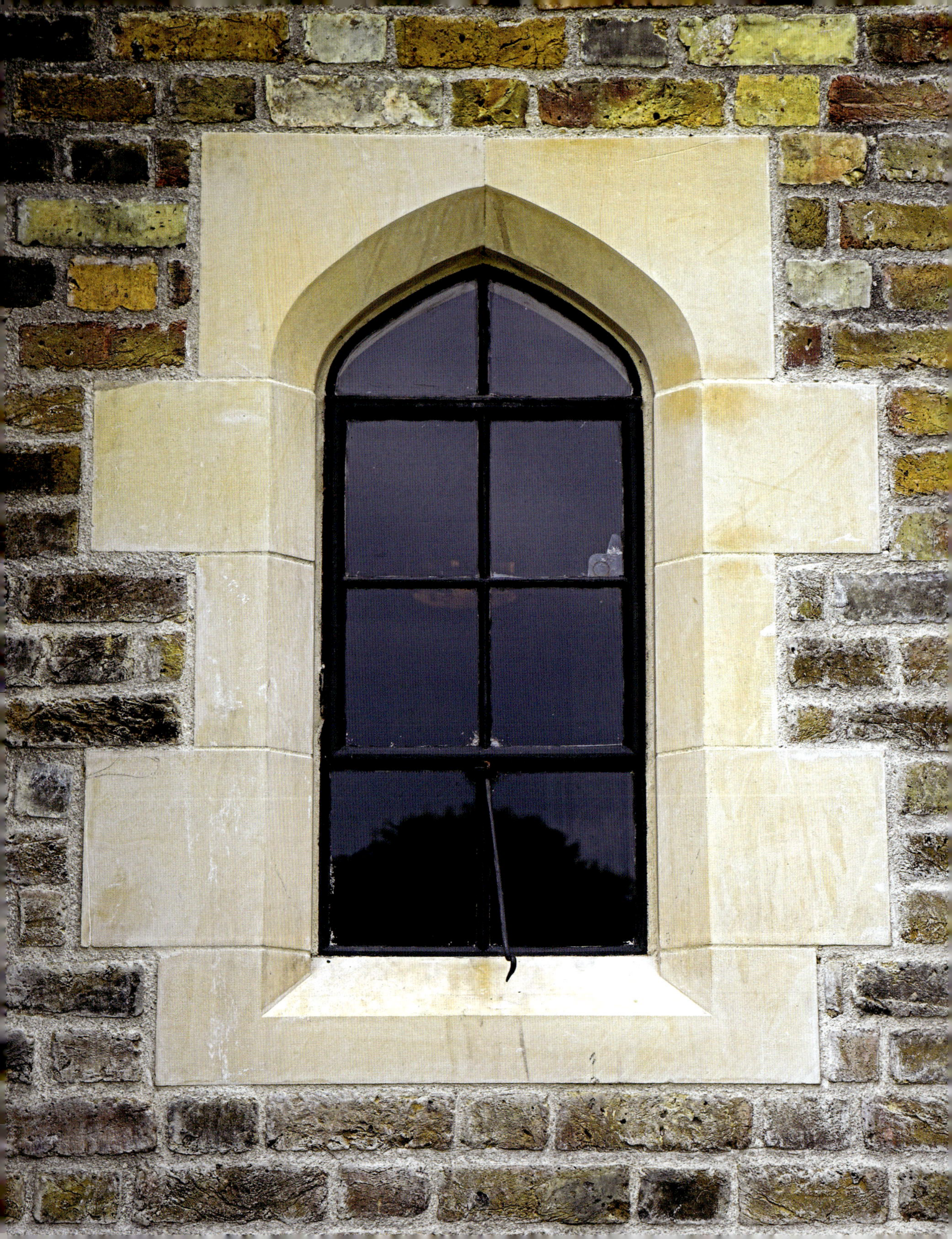

REPAIRS

this time suggested 15mm or a minimum of twice the joint width, whichever was greater. Today, specification 'inflation' has resulted in us now being required to rake out joints to a minimum depth of 30mm. Whether this is strictly necessary is a matter of judgment, but it should be noted that costs will obviously increase with the requirement for depth.

Raking out a joint to a specified depth needs considerable skill to avoid damage of the stone or brick. Power tools should always be avoided unless they are appropriate for the task and the requisite skills are available. In many cases, power tools are unnecessary as the mortars are sufficiently weak to be removed without risk of damage. If power tools are required to remove mortars, it is sometimes more appropriate to leave the existing mortars in place unless it is evident that the masonry is being damaged by their presence.

It is helpful to analyse the original mortars so that the sands and aggregates can be identified and matched. Most analysis is undertaken by dissolving out the lime binders using acids, but this will also dissolve out any limestone dust that was originally incorporated into the mix (a problem identified by one of our particularly bright conservation graduates). This important ingredient is therefore often overlooked. In our opinion the use of stone dust helps a lime mortar as it seems to retain moisture, aid adhesion, and achieve a more controlled carbonation.

Typical mortars will be based around a lime putty or naturally hydraulic lime of which there are three grades: feeble, moderate, and eminently strong. A general starting rule of thumb is that there should be approximately three parts of well graded stone dusts, sands, and crushed stone aggregates to each part of lime. The permeability and strength of the mortar must be selected to suit the masonry.

Masonry joints should be wetted prior to the application of new mortar so that the masonry does not draw out water from the mortar before the reactions have occurred. In hot weather, the new mortar should be protected from sunlight with hessian, and sprayed regularly to ensure that suitable humidity is maintained. In cold weather, the mortar needs to be protected from frost using polythene drapes.

New pointing work should be protected from the rain.

The colour and texture of the new pointing needs to be carefully considered. It is often desirable to rub back the mortar with hessian after the initial setting period to expose the aggregate.

Equally important is the style in which the mortar is finished. The use of a steel pointing tool to achieve a weather-struck joint is never appropriate on an

LEFT
Caen stone replacement window details and repaired brickwork at August Pugin's home The Grange, Ramsgate

ABOVE
Brickwork and stone prior to repair at August Pugin's home The Grange, Ramsgate

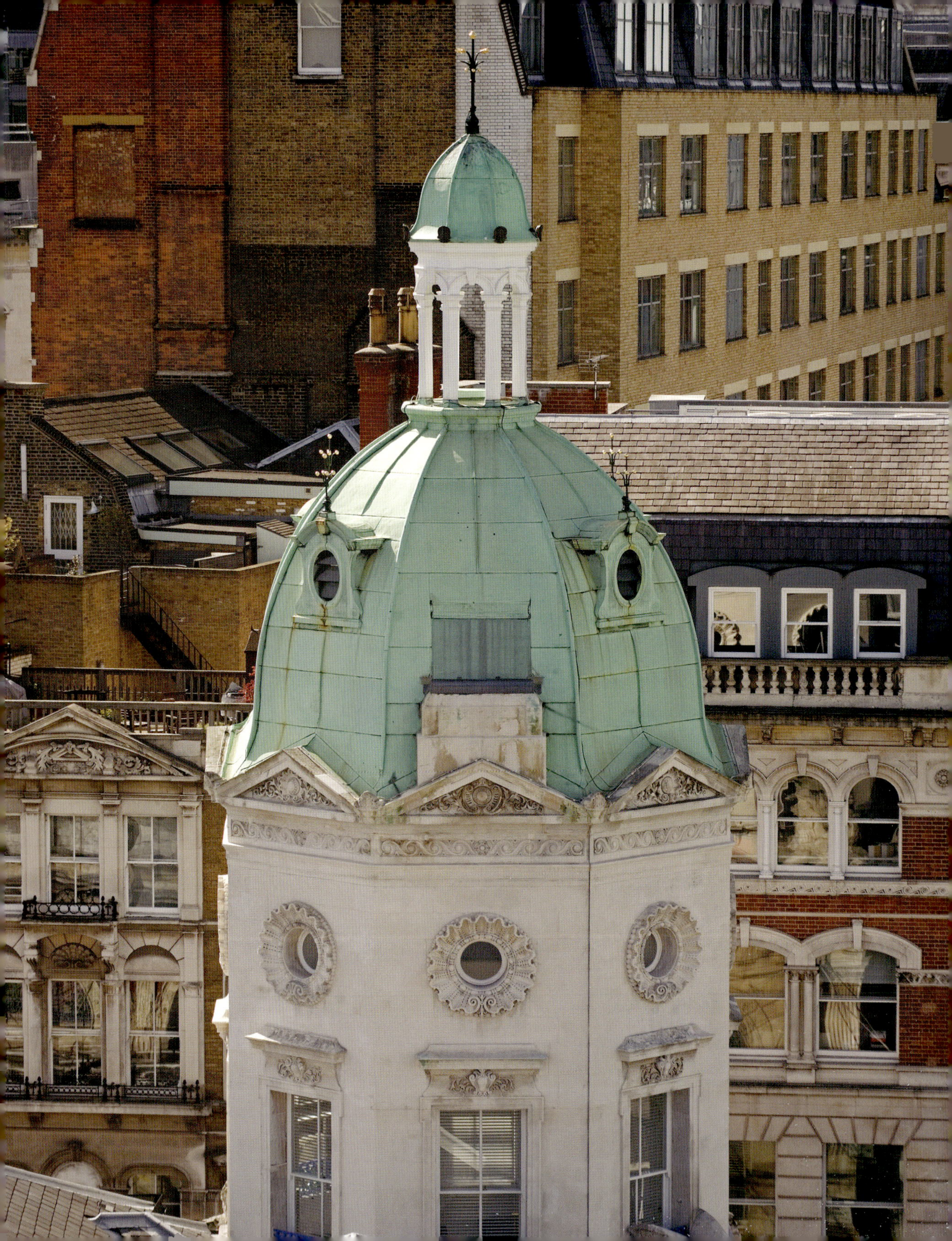

REPAIRS

historic façade. It is generally considered appropriate to lightly expose the arrises of the bricks or stones.

Samples should be presented for approval before commencing repointing works.

REDRESSING (RE-WORKING) STONE

Where a stone is degrading and forming hollow cavities which have the potential to hold rainwater and become saturated, it is sometimes appropriate to carefully and sensitively re-work the stone so that the water traps are removed. In this way the original stone can be retained at least until the next maintenance works programme. Ultimately, the stone may need to be replaced.

Aesthetics will also need to be considered as one degrading stone can spoil the appearance of an otherwise well maintained façade.

Where a carved stone is weathering and losing its definition, it can sometimes be appropriate to rework the stone to redefine the detail. However, this technique should not be used if original tooling marks are lost as a result. This is very subjective and careful consideration is always necessary.

REPLACING STONE

The decision to replace a stone on an historic façade can be a complex process that will reflect the needs of the building, its owner, and the budget. PAYE have extensive experience in this area and are often asked to provide technical and cost advice.

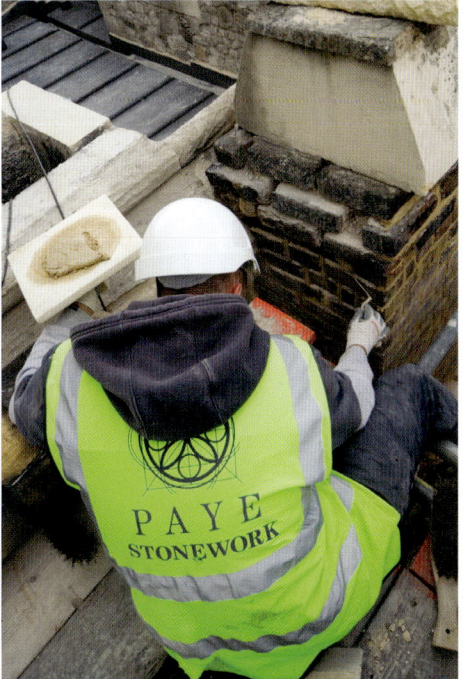

ABOVE
The Beauchamp Tower at the Tower of London after conservation

LEFT
Parapet wall repairs at the Beauchamp Tower, the Tower of London

FAR LEFT
Portland Stone repairs Smithfield Market

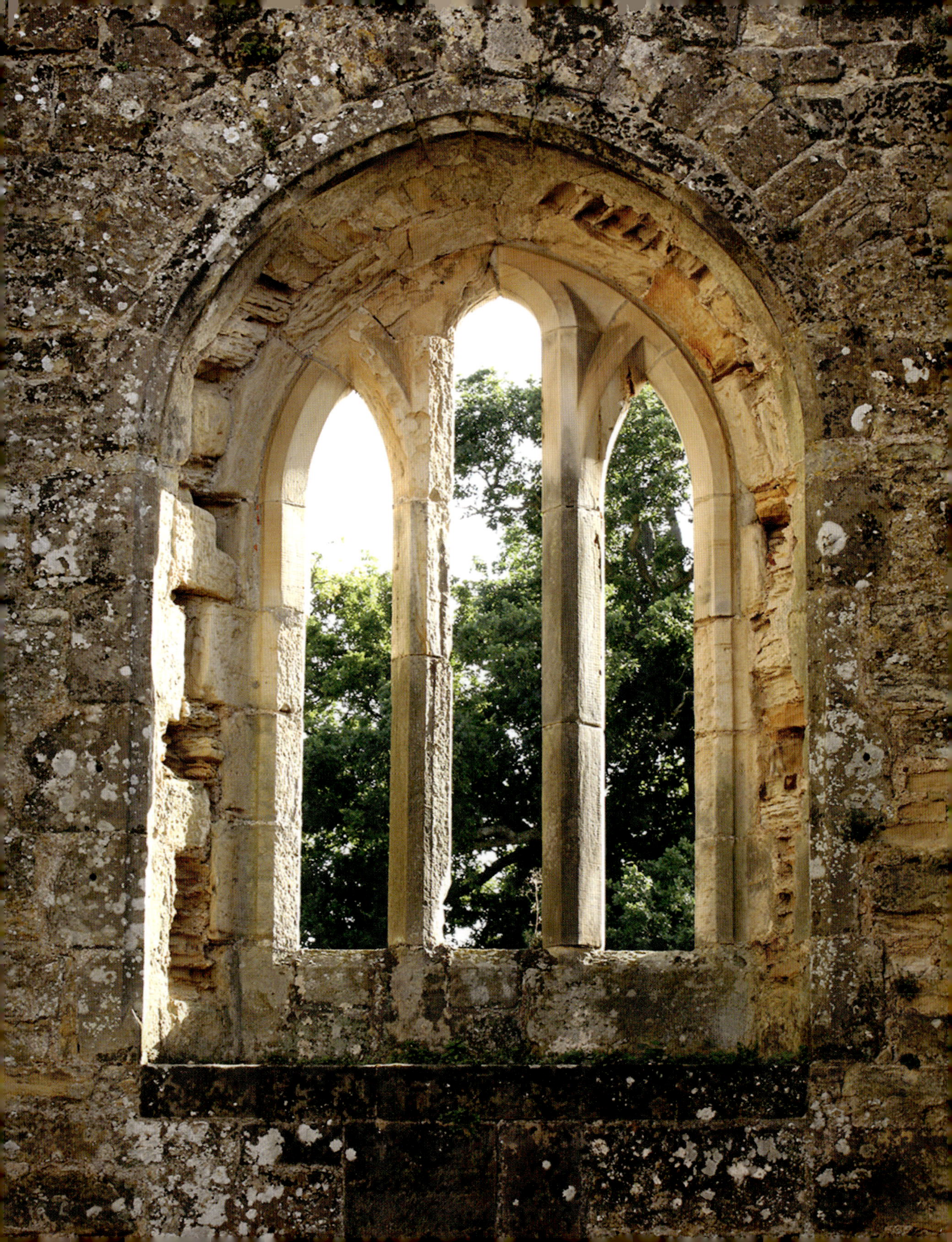

REPAIRS

An owner, occupier, or developer will often be primarily concerned with aesthetics and cost. The needs of the building will relate to the rate of degradation, the planned maintenance period, the management of rainwater, and the need for mitigation works. In principle, stone should only be replaced when necessary and when other repair methods do not economically provide suitable long term solutions. The conservation of original materials and safety are usually the most important considerations.

The sourcing of suitable replacement stone is an important first step. It is usually possible for PAYE to visually identify the original type of stone and recommend suitable quarries and beds of stone. Many of the original quarries have been worked out and closed. Other quarries may still operate but the current work face might be producing stone of different characteristics to the original. Therefore a pragmatic approach to the approval of suitable samples is required. However, the samples must be realistic and present the various characteristics that should reasonably be anticipated by the quarry operators. If too exacting criteria are applied in the approval of samples, costs will escalate in line with the increased percentage of waste. It is appropriate to be more selective for masonry at lower levels where certain characteristics will be more visually apparent.

It is also appropriate to respect the original joint layouts and joint widths. Introducing smaller stones onto a façade might reduce repair costs but adversely affect the appearance, particularly at lower levels.

Replacement stones should be fully bedded and grouted in position, and physically restrained using non-ferrous fixings as appropriate.

Some conservationists prefer the replacement stones to be apparent as such in what they regard as an 'honest' repair. Other conservationists require the repairs to blend imperceptibly into the existing masonry.

Often, we are requested to install the front profile of the replacement stone in the original façade plane so that it stands proud of the adjacent weathered stones. Other conservationists require the stone to be set in line with adjacent stones and be lightly dressed to replicate the long term effects of weathering.

It is not always necessary to replace complete stones. If carefully considered, it is often possible to discretely introduce

FAR LEFT
Bath stone indents at Bodiam Castle

ABOVE
Bath stone indent repairs, Windsor Castle

additional new joints by only replacing part of a stone: a technique referred to as indenting. It can be suggested that from a conservationist view, it will always be preferable to retain as much of the original material as possible in this manner.

MORTAR REPAIRS

English Heritage stopped giving grants for mortar repairs on stone buildings over 20 years ago, the reason being that these repairs were being used indiscriminately without sufficient care and attention, and typically used cement mortars which were often modified with inappropriate PVA additives to improve adhesion and increase the waterproofing abilities of these mortars. Not only did these repairs fail in the short term, they caused the underlying stone substrate to decay at a faster rate than it would have if left unrepaired. At best they wasted the client's budget, at worst they caused further damage to the stone and increased overall long term maintenance costs.

The use of mortars to repair the surface of stone buildings became more common after the Second World War. Mortar repairs are much cheaper than replacement stone, but rarely provide economic advantage into the long term. Lime mortars fare better than cement mortars particularly if only used in locations which are not susceptible to saturation from rainwater. Cement mortars also cause the stone substrate to deteriorate.

Mortar repairs are still common in Europe, perhaps because they do not suffer the repeated wetting/drying and freezing/thawing cycles from which we suffer in the UK. Europe is where many of the repair mortars now sold in the UK are sourced.

In our opinion, care needs to be taken when using a proprietary product derived from the concrete repair industry rather than a more appropriate bespoke lime mortar.

If mortar repairs are to be used, they must be more permeable than the stone substrate and sufficiently keyed. If necessary, the repair should be reinforced with non-ferrous dowels. Mortar repairs should never span across a masonry joint.

Typical repair mortars will feature a lime putty or naturally hydraulic lime of which there are three grades: feeble, moderate, and eminently strong. A general starting rule of thumb is that there should be approximately three parts of well graded stone dusts, sands, and crushed stone aggregates to each part of lime. A repair mortar must be less permeable than the natural stone substrate.

The colour and texture of a mortar can be varied to replicate that of the natural

ABOVE
St Mary Virgin, Reading

RIGHT
Saint Matthias Church, Richmond

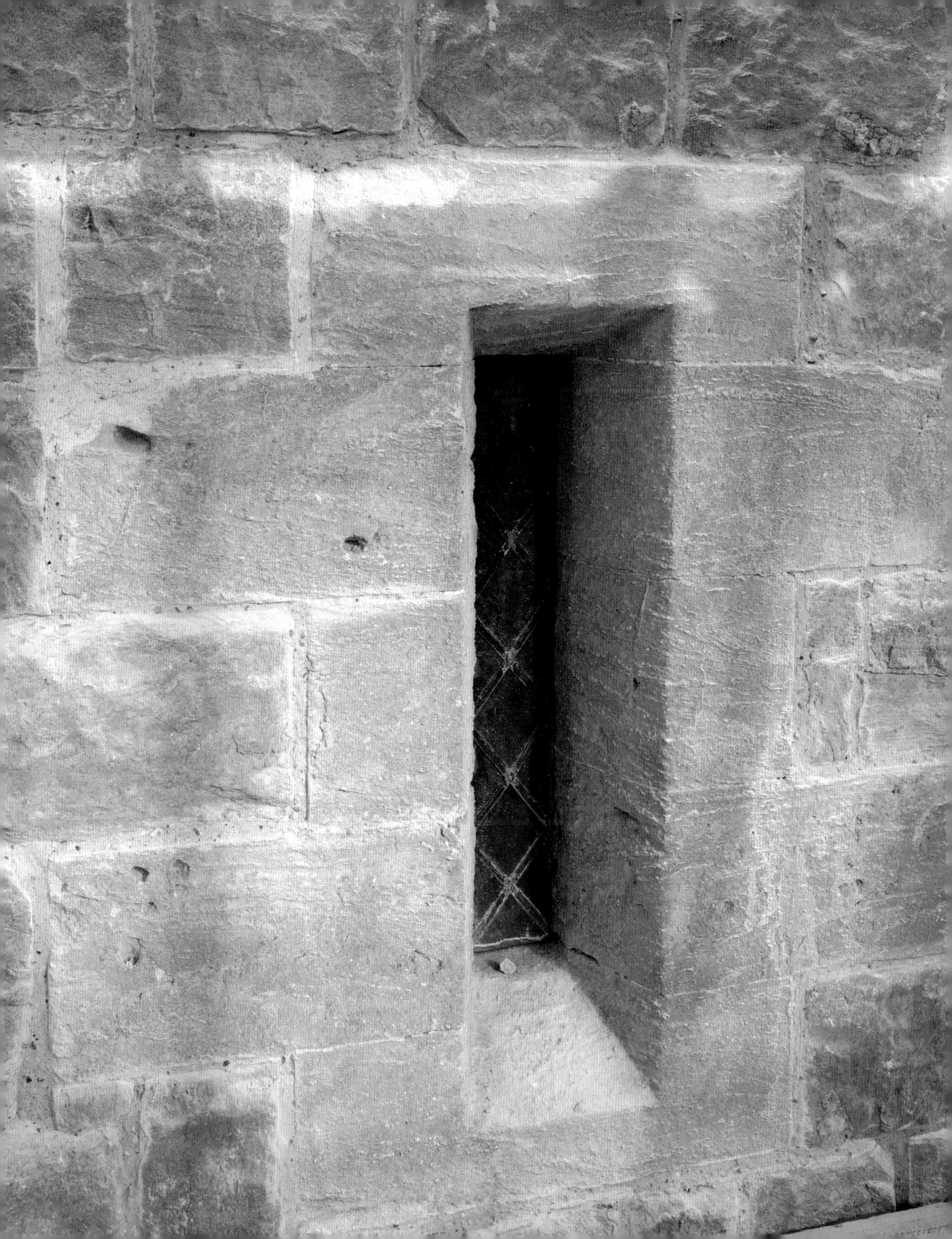

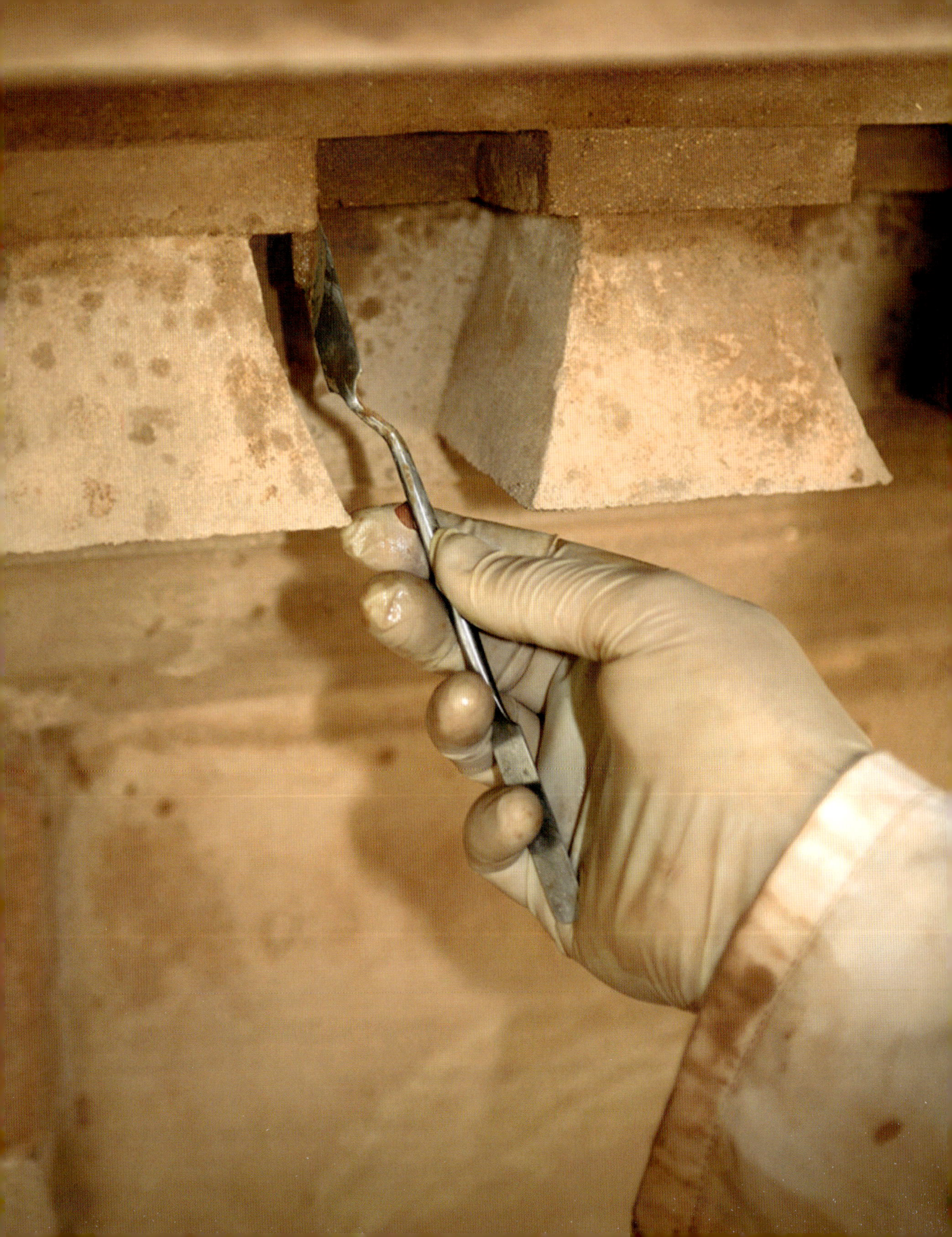

stone by using different stone dusts, crushed aggregates and sands. It is sometimes appropriate to finish the repair proud of the adjacent stone surface and lightly work back the surface to expose the aggregate and provide a suitable texture.

SACRIFICIAL SHELTER COATS

Where a stone is degrading but it is not deemed appropriate to replace it, a shelter coat can be applied over its exposed surface to help protect that stone from further atmospheric attack. The principle is that decay subsequently occurs within the shelter coat which will eventually fail and need to be replaced. In this way, the stone below is protected.

Shelter coats must be more permeable than the stone substrate. A typical mortar mix would involve a weak lime putty and finely sieved stone dust and sand.

Shelter coats are often used in marine environments where the stone is regularly subjected to wind sprayed sea salt crystallisation damage.

STITCHING FRACTURES

This common and useful technique involves installing non-parallel pairs of strengthening rods across the fracture. Setting the rods at an angle to each other ensures that the fractured portions will not be able to pull apart.

For the last 30 years or so, stitching rods have been threaded stainless steel set in resin. The bars must be threaded to ensure good adhesion to the mortar or resin, and are typically 4, 6 or 8mm diameter. It is important to blow out the dust from the drilled hole before the mortar or resin is inserted to ensure that a good bond is achieved with the stone.

An alternative to the threaded rod is a twisted helical plate. This is less stiff than a rod, but is preferred by some consultants because it does facilitate a degree of movement. Being of a much smaller section than the threaded rod, much greater volumes of resin are required to fill the drilled holes.

At PAYE we have have carried out trials using basalt fibre rods as an alternative to stainless steel, and setting these in hydraulic lime mortar. Early results are encouraging and we anticipate their widespread use in future years. The masons also prefer to use them for mouse-dowelling because they can form a sharpened point at one end to easier locate the dowel into its mortice. Furthermore, there is much less embedded energy in basalt fibre than stainless steel and more compatibility to stone in terms of thermal dynamic performance.

Stitching a fractured stone should be

LEFT
Lime mortar repairs in protected locations, London Hippodrome

ABOVE
Conservation repairs in progress, London Hippodrome

considered as a repair which treats the symptom but not the cause. To effect a long term repair it is necessary to understand why the stone has fractured and to remediate the cause as well as stabilising the fractured stone otherwise another fracture will form, usually parallel and close to the original.

Causes of fracturing could include the failure of a backing lintel, differential foundation movement, thermal expansion, or the corrosion of embedded metalwork.

If the fracture is wide enough, it should be grouted with an appropriate lime mortar. Narrow fractures can be resin injected if deemed necessary, but if no movement has occurred it is sometimes more appropriate to leave the fracture unfilled.

CONSOLIDATION OF FRIABLE SURFACES

For centuries products have been sought which have the ability to stop stone decaying and to strengthen a failing stone. In consequence, many different solutions and products have been applied to failing masonry surfaces. Most have been found to damage the stone though in recent years a few have been developed which appear to provide positive benefits. Examples of previous, sometimes disastrous, surface treatments include wax, paraffin, linseed oil, shellac, paint, polyurethane, varnish, and silicon. Any application which traps moisture will cause the stone substrate to degrade. It is often considered preferable to lightly defrasse the masonry surface to carefully remove delaminating material and leave it to slowly degrade until the next scheduled repair programme of works.

It is sometimes appropriate to apply a consolidant in order to introduce a degree of cohesive integrity. A number of consolidation techniques are available.

Limewatering is the traditional approach for limestones and involves 20 to 40 applications of water saturated with calcium carbonate. The solution is absorbed before the water evaporates, depositing the lime in the masonry pore structure.

Siloxanes (Silanes) are supplied in water repellent or non-water repellant applications and applied 'wet on wet' to encourage the solution to penetrate as deeply as possible into the masonry pore structure. The first UK application of these was in the 1980's and to date does not appear to have caused any degradation of the Bath stone and Portland stone

ABOVE
Kentish Ragstone indents and grouting works, Tower of London

RIGHT
Saint Matthias Church, Richmond

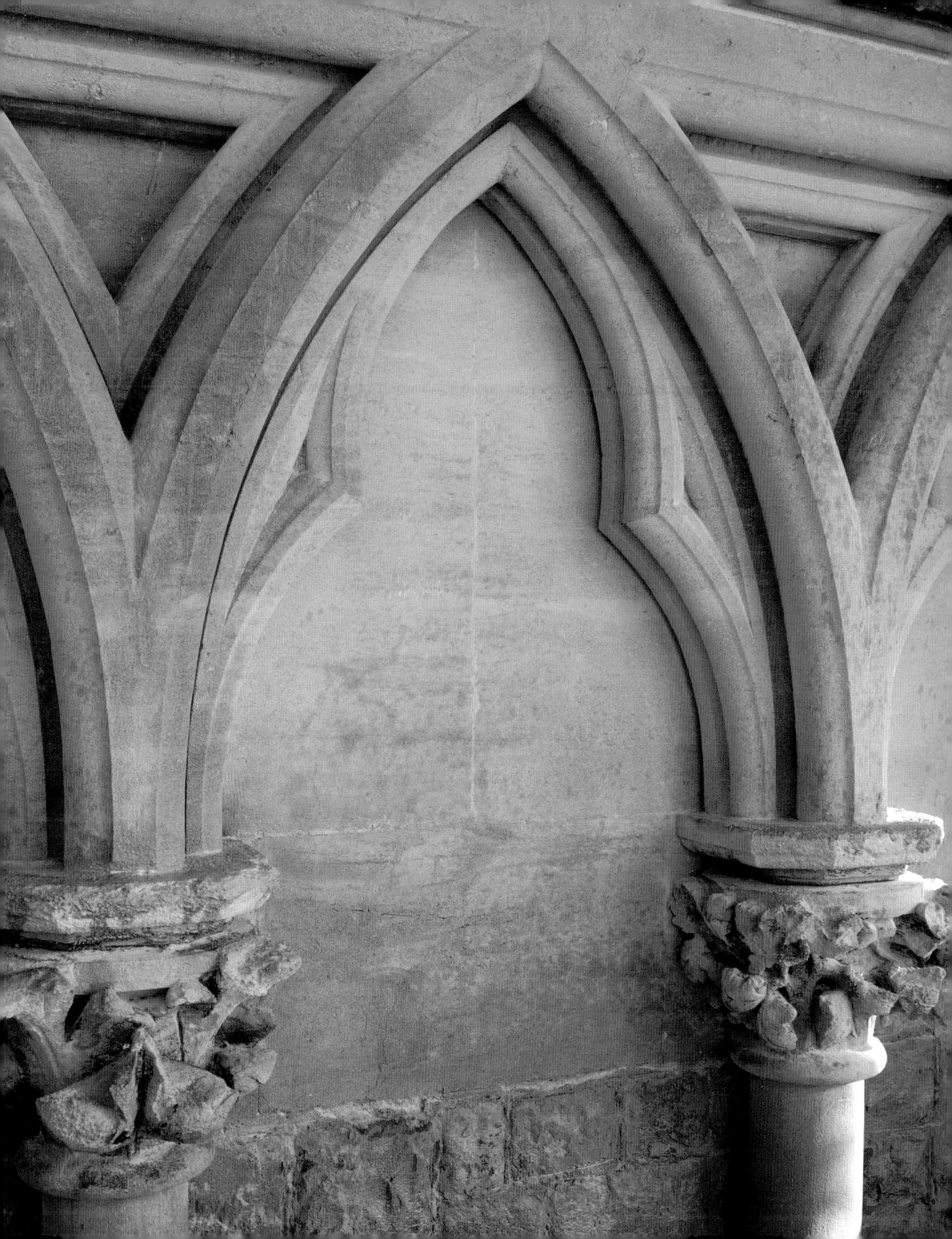

The reconstruction of the crypt vault, Darnley Mausoleum

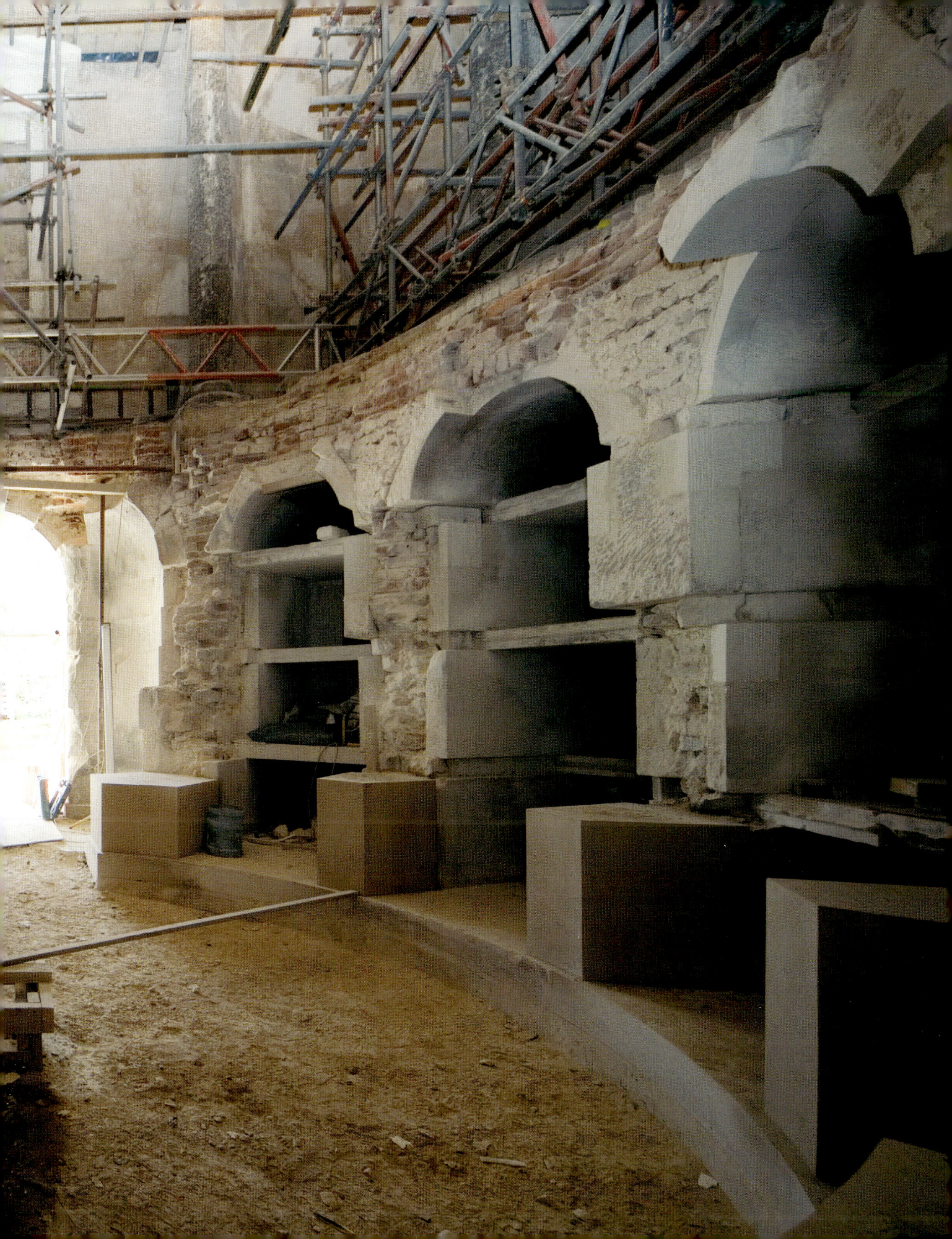

LONDON HIPPODROME

PAYE
Stonework & Restoration
020 8857 9111

REPAIRS

substrates. These products were often sold on the manufacturer's promise that once treated, a masonry façade would never again need cleaning. As a result some complete façades were treated but, perhaps predictably, gathered the dirt and became soiled as they would had the siloxanes not been applied.

Nano limes involve the application of quick limes in ethanol solutions which are thought to achieve better penetration and consolidation than the above techniques.

Waterproofing products from DIY stores should be avoided.

If consolidation is deemed to be necessary, we would suggest that it is always preferable to treat individual stones rather than complete areas.

LEFT
Cleaned and repaired stone façades with new terra cotta centurian, London Hippodrome

ABOVE
Drawings of reredos repairs at the Darnley Mausoleum

87

Conservation work at Windsor Castle

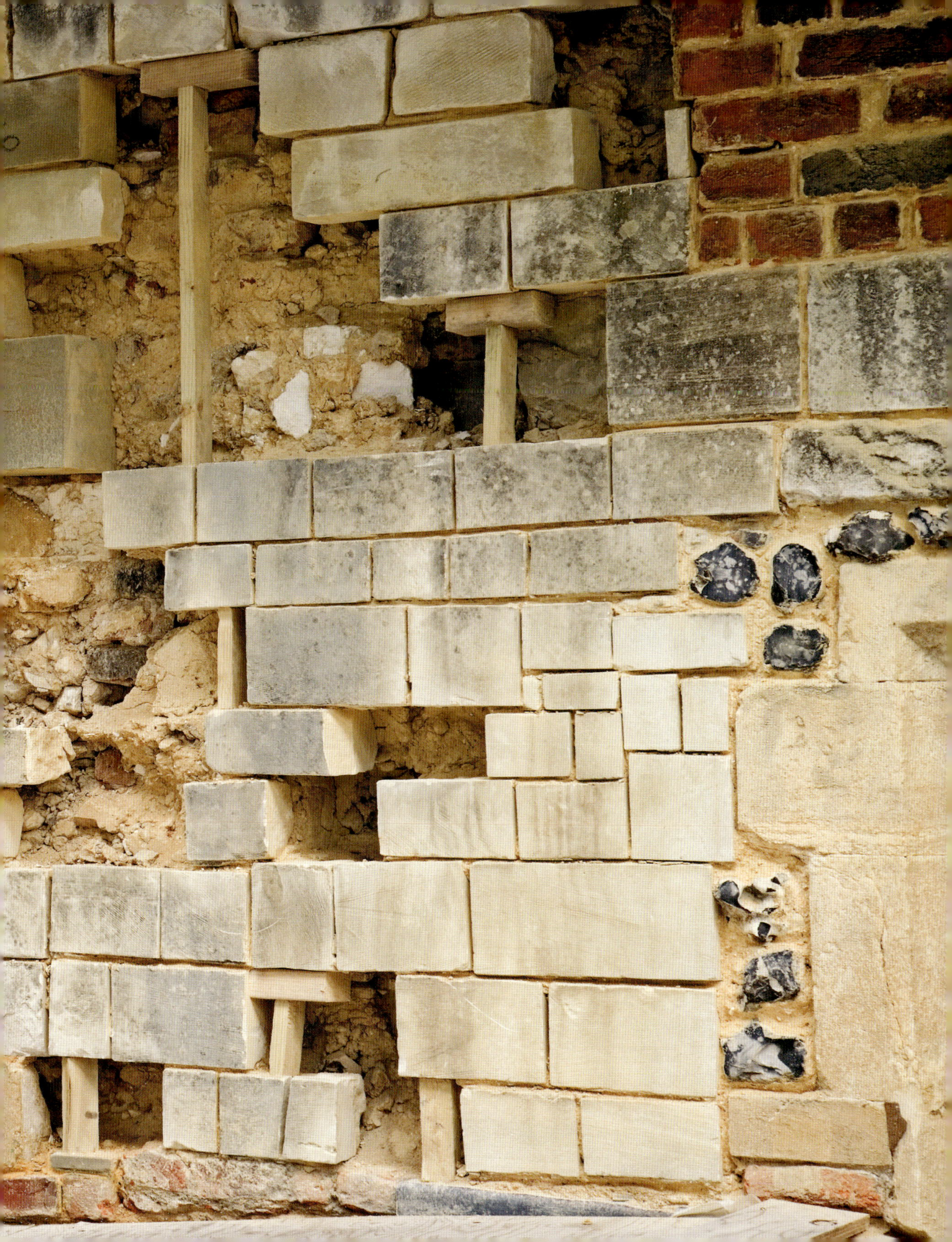

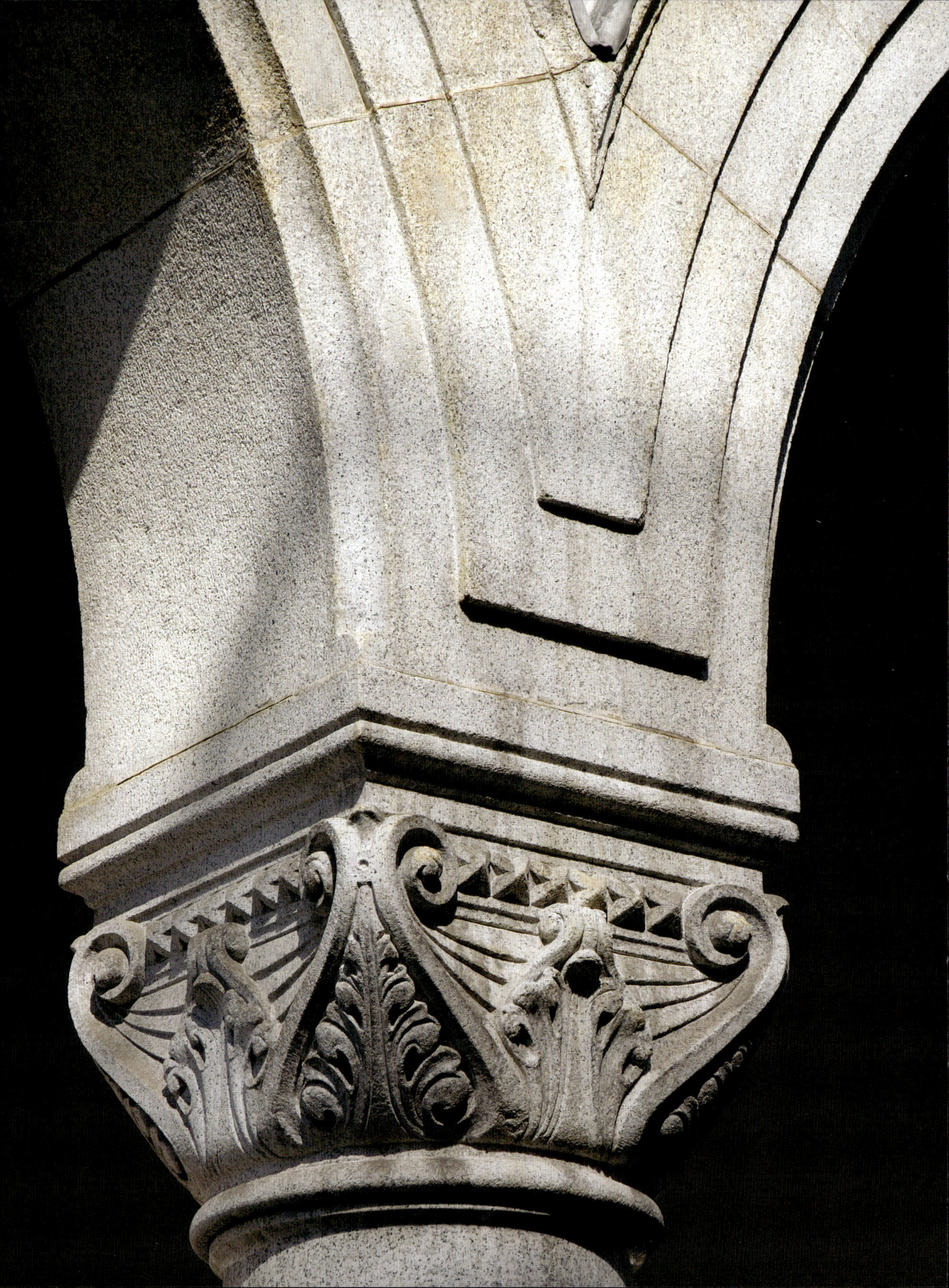

CHAPTER FIVE

TECHNICAL ISSUES, MANAGEMENT & ORGANISATION

This chapter describes the many technical issues which should be considered in effectively managing a façade repair works programme.

STRENGTH IN COMPRESSION

Stone is immensely strong in compression but relatively weak in tension and shear. Traditional loadbearing masonry must work in compression or be designed in such a way that imposed tension, shear and torsional loads are never more than nominal. Failure to do this will result in the stone cracking. The stone is so weak in resisting loads other than compression that often only changing a load path will result in fracturing.

The Romans developed the use of the arch where the individual voussoir stones work in compression to transfer imposed loads around the opening. A plain stone lintel above an opening can only span a limited distance. By using arches in lieu of lintels, the Romans were able to build enormous aquaducts, viaducts, and other major engineering structures.

Granite capital detail at Cadogan Hall

The arch was significantly developed in medieval times when it evolved a point. This in turn lead to the development of vaults, flying buttresses and tracery windows: all with masonry units working in compression.

ORIENTATION OF BEDDING PLANES

It is important to correctly detail and orientate each stone so that the loads are imposed perpendicularly to the bedding planes in the same conditions that the stone formed in the ground, compressed by the mass of material above for millions of years. In this way, the compressive strength of the masonry is maximised.

Good practice also dictates that to maximise durability, the bedding planes must not be orientated parallel to the exposed surface. Failure to do this can result in the de-bonding of the bedding planes close to the surface causing laminates to peel away from the substrate below.

The orientation of bedding planes is more important in some stones than others. Many sandstones have pronounced bedding planes so it is important to use these stones in their correct orientation.

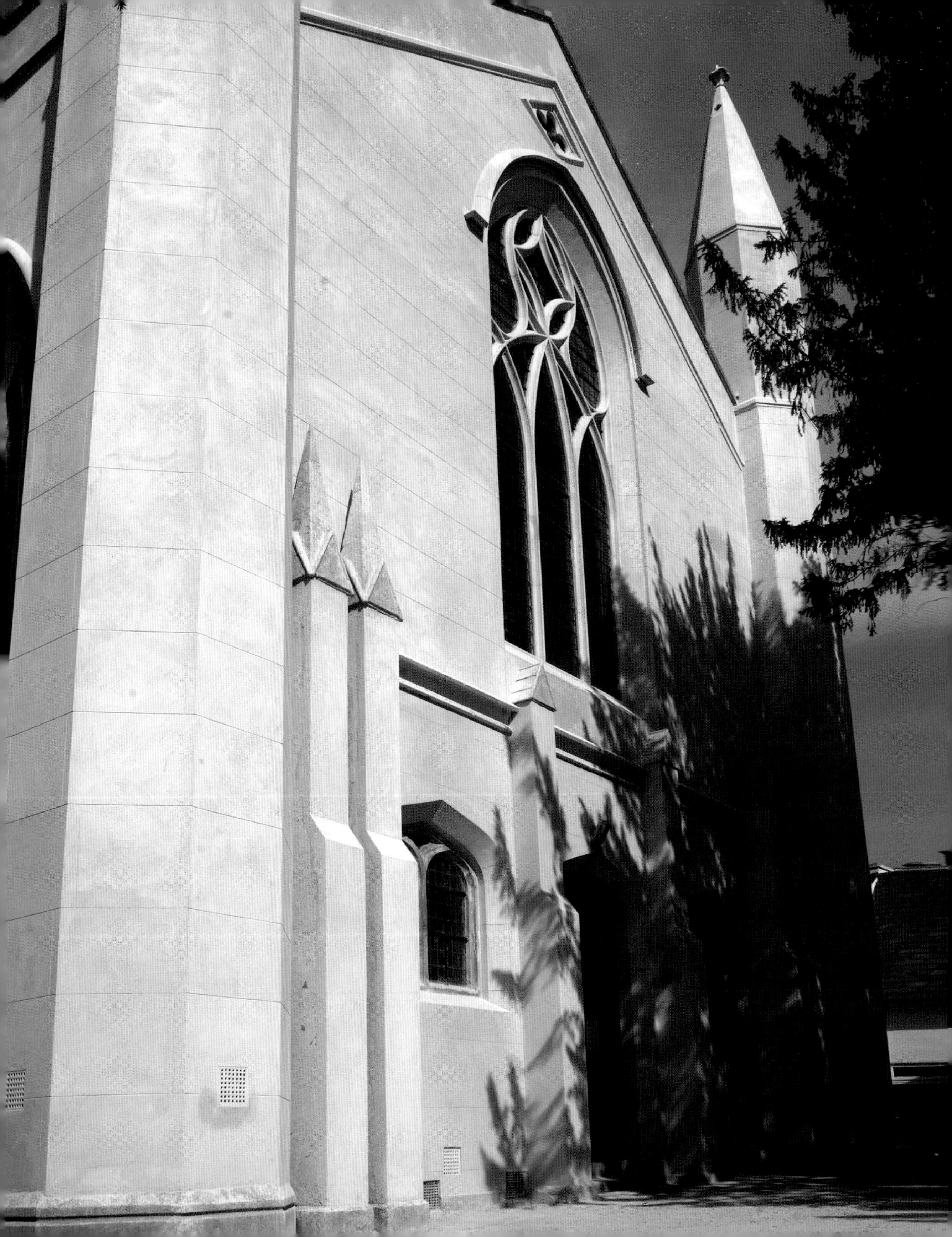

TECHNICAL ISSUES, MANAGEMENT & ORGANISATION

Projecting units such as cornices are orientated with their bedding planes vertical and set perpendicular to the supporting masonry below in order to provide the maximum strength of cantilever.

SELECTION

In the UK we have unfortunately witnessed some spectacular failures by using imported stones which are unsuited to our particular climate of repeated wetting/drying and freezing/thawing cycles. This is not to say that all UK stones are suitable for use in any location, but a particular stone's ability to withstand the adverse effects of the weather is more likely to be known if that stone has had a long and continuous use.

Kentish Ragstone was used by the Romans to build their city of Londinium nearly two thousand years ago. Portland stone has been used in London since the 17th century, and since the construction of the Kennet and Avon canal made it economic to transport Bath stone in the early 19th century, Bath stone has been frequently used in London. The railways made it possible to transport Red Mansfield and Bramley Fall stones economically into London from the mid-19th century, and York stone paving has been used in London for centuries. Whilst recognising that the quality of the stone may well change over the decades as old quarries become worked out and new quarries open, these stones are effectively tried and tested. Their durability is known. It is advisable to view a stone that has been used in its intended way for many decades or centuries.

A stone that might be suitable for use as protected ashlar walling might not be suitable in an exposed location such as a coping, cill or paving stone.

Laboratory testing of stone samples can replicate the adverse effects of the weather and measure compressive and flexural strengths, and these results can provide indications of how an untried stone might perform in service. However, extreme care is needed to correctly interpret these results. Mistakes will be costly. For instance, tests are undertaken on small pieces of stone, often only 40mm cubes. The chances of one cube achieving identical results to any other are small as the characteristics of many stones can vary considerably across a single quarried block. Another problem is that different laboratories might have slightly different techniques, which could result in two different laboratories achieving different results from a theoretically identical cube. It is generally best to ensure that one laboratory undertakes the testing for a prospective stone against an indicative range of samples, and that similar tests are undertaken concurrently on a known stone such as Portland. This way, a realistic comparison will be achieved.

It is best to be cautious when assessing any results provided by quarries. Though most of the UK quarries are operated by

LEFT
Portland stone façade after cleaning and repair, St Paul's Church, Mill Hill

ABOVE
Portland stone steps, East India Club, St James's Square

TECHNICAL ISSUES, MANAGEMENT & ORGANISATION

professional companies who will be keen to ensure that their stones are used appropriately, there are many instances where inappropriate materials have been sold to trusting clients. In the late 1980's Portland stone was in great demand.

Costs were rising steeply and certain companies attempted to gain commercial advantage by selling imported stones which were deemed to be 'similar'. One such stone was imported from eastern Europe. Its technical data was provided by the UK company selling it, and it was consequently specified on a number of new-build projects. However, the durability of that stone was so poor that coping stones were failing even before the scaffolding was struck. The importing company had commissioned a series of laboratory tests and repeated these until, eventually, a few cubes managed to achieve satisfactory results. These results were not at all indicative of the durability of the stone but were nevertheless used to market it.

SCHEDULING / REFERENCES / DRAWINGS

Ragstone and rubble walling use stones which are generally supplied direct from the quarry in small sizes that are easily man-handled and only nominally worked as the masonry is constructed. This is a comparatively economic walling system and was used extensively from prehistoric times. The procurement of such stone today does not need detailed drawings. The most common type of stone used in this way in London, was Kentish Ragstone due to the fact that this was the nearest source of durable stone that could be transported by barge (water transport being the most economical means).

RIGHT
Repairs in progress at Kenwood House for English Heritage

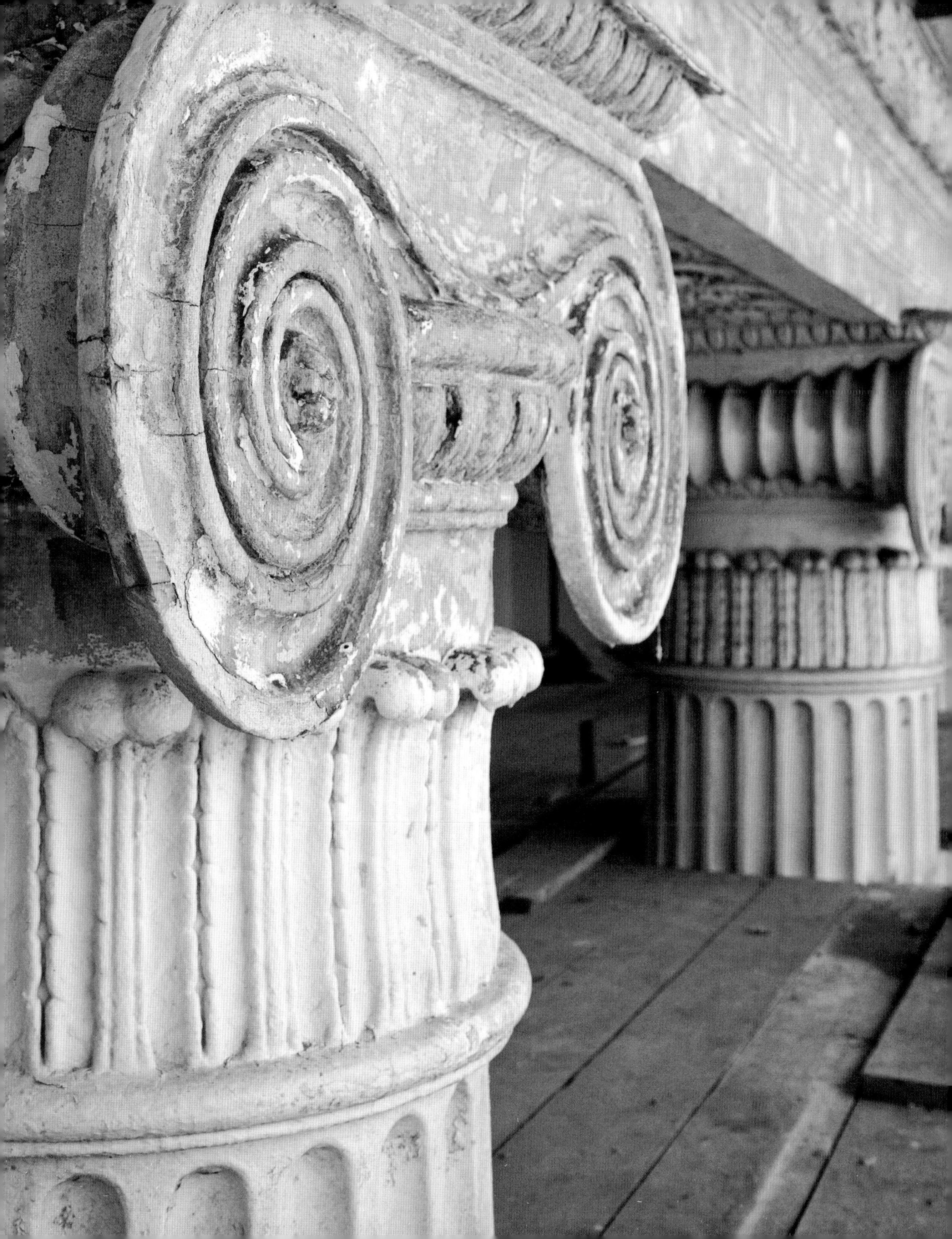

TECHNICAL ISSUES, MANAGEMENT & ORGANISATION

Above
Dismantling the Aston Webb screen wall at the V&A Museum

Right
Dismantled blocking course showing original cement grouted joggle joints

Worked dressed stone is individually cut to accurate dimensions. Attention to detail is the key to successful control, timely delivery, and proper installation. If an existing façade is being adapted or extended, it is necessary to undertake detailed and accurate dimensional surveys. Elevations and plans are required to show individual stones, joint widths, course heights, control dimensions and joint patterns. From these, individual stone sizes can be determined. Each stone is given an individual reference number relating to the elevation, course number and location on that course. Conventionally, individual referencing numbers run from the lowest course upwards and from left to right ie. generally in the sequence that the stones will be laid. It is important to give individual stones individual reference numbers to assist the masons installing the units on site.

In the workshop, cutting schedules are prepared for the banker masons to work each individual stone involving a 3D sketch showing all necessary dimensions and workings. From these sketches the stones are cut. Finished stones are packed on pallets or crates to suit the handling conditions on site, taking care to protect arrises with non-compressible and non-staining spacers.

Increasingly, stones are being cut by computers direct from CAD drawings, however it is usually necessary to hand finish these units to correct profiles and achieve an appropriate tooled finish.

The stones must be packed, crated, and transported to site in the correct installation sequence: there is rarely enough space on modern construction sites for any unnecessary storage. Any stone delivered out of sequence will become a liability, potentially moved many times and therefore more prone to damage. Accurate schedules

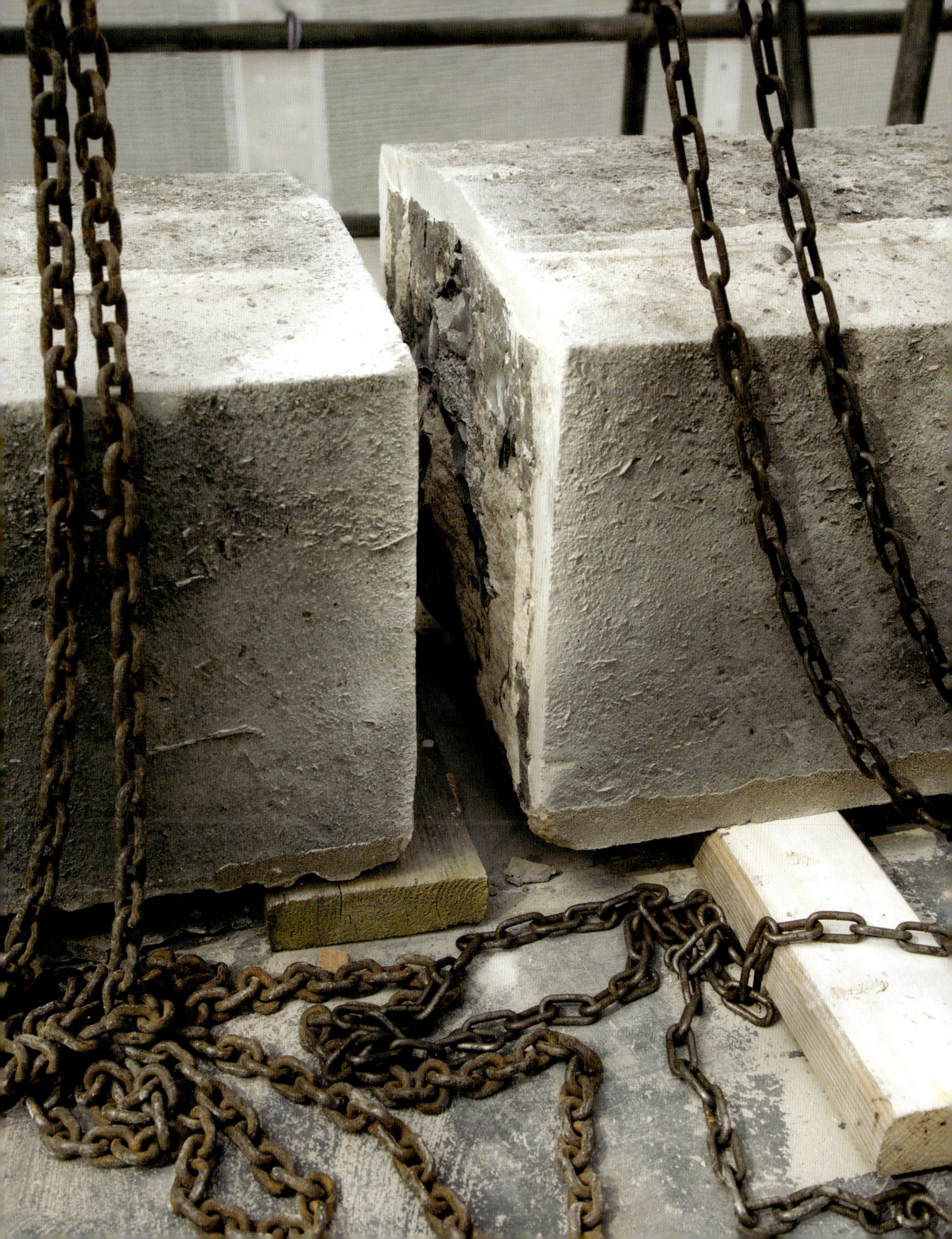

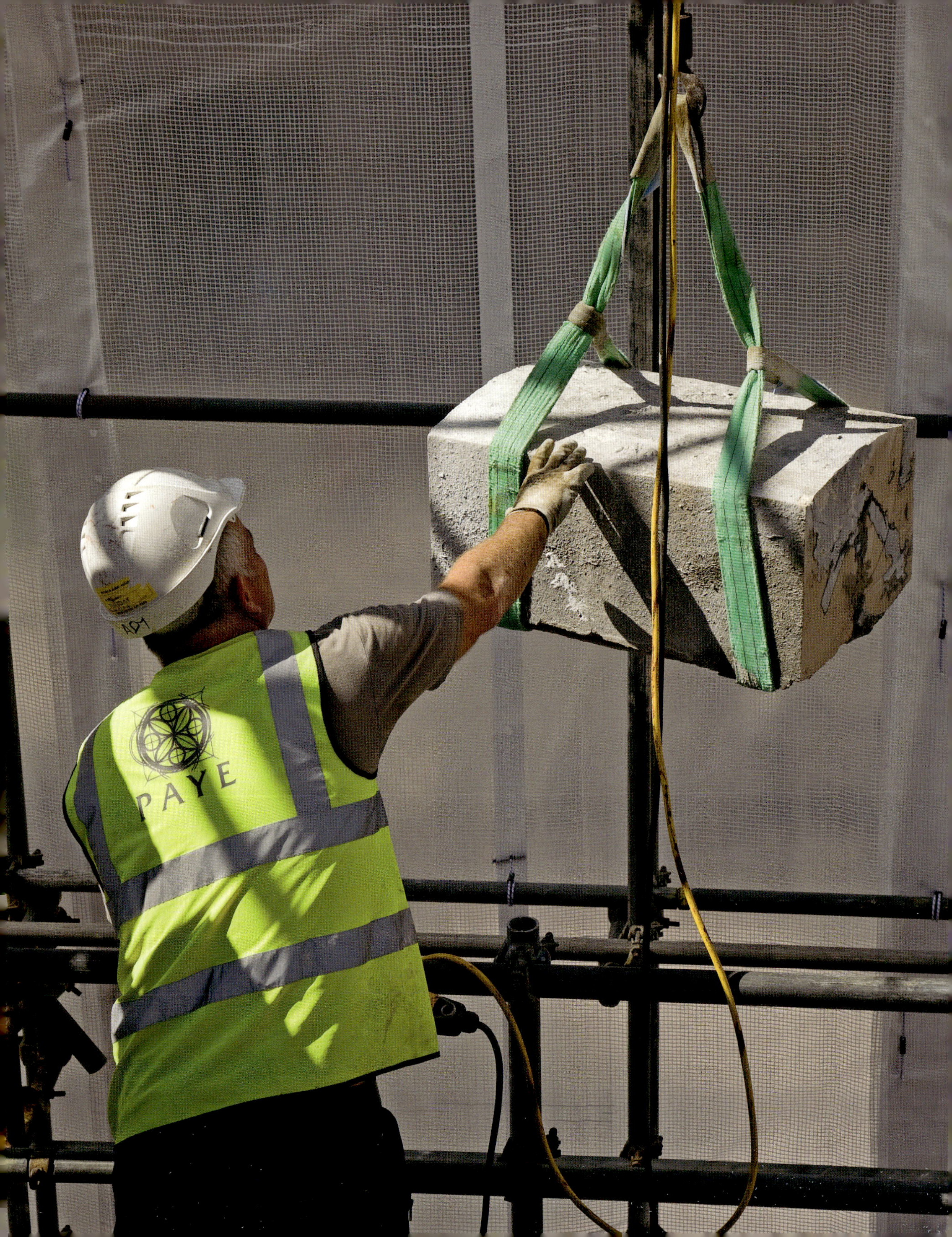
PAYE

TECHNICAL ISSUES, MANAGEMENT & ORGANISATION

must be kept recording the reference numbers of individual stones packed into numbered pallets to facilitate the efficient transport, unloading, hoisting, distribution and installation works on site.

UNLOADING, DISTRIBUTION, AND INSTALLATION

The health, safety and welfare of the individuals involved at all stages of the production process are of paramount importance. The unloading, distribution and installation of the crated and un-packed stone poses risks which must be effectively managed and eliminated. Health and safety must never be regarded as 'red tape'. It is an essential and necessary part of every process.

Stone is heavy. A loaded pallet or crate can easily weigh more than a tonne and all packaging and lifting equipment must be designed and certified to be suitable for the purposes intended. The density of a limestone, which is often still wet from the sawing operations, can be as much as 2.7 tonnes per m^3.

Therefore a stone of size 700x400x200mm will weigh 151kgs, which is much more than two masons can safely lift without mechanical handling equipment. The maximum weight which can be lifted safely is never defined because it depends on the abilities of the specific individuals involved. A useful guide is that any stone weighing more 50kgs will need to be unloaded, distributed and installed by means of certified mechanical lifting equipment. Masons fixing stone need to be trained to lift loads safely. There is much potential for injuries to backs, feet, wrists, eyes and fingers.

PERSONAL PROTECTION EQUIPMENT (PPE)

PPE includes as a minimum, safety helmets, steel toe-capped boots, gloves, appropriate eye goggles, and high visibility clothing. Should a particular operation need additional PPE, this must be identified as part of the risk assessment and be defined in the method statement. It will be necessary for appropriate breathing masks to be worn by masons undertaking work which creates dust. Sandstone dust is the most dangerous because it contains silica.

It is essential for the individuals concerned to be properly briefed by means of tool-box talks to ensure that they fully understand the working methods and the purpose of the safety equipment provided. We consider it important for all management, including consultants, to set an example and always wear a similar amount of PPE whilst on site in work areas.

LEFT
Carefully lowering a dismantled blocking course stone for transport to storage facility

ABOVE
Palletised stonework in storage

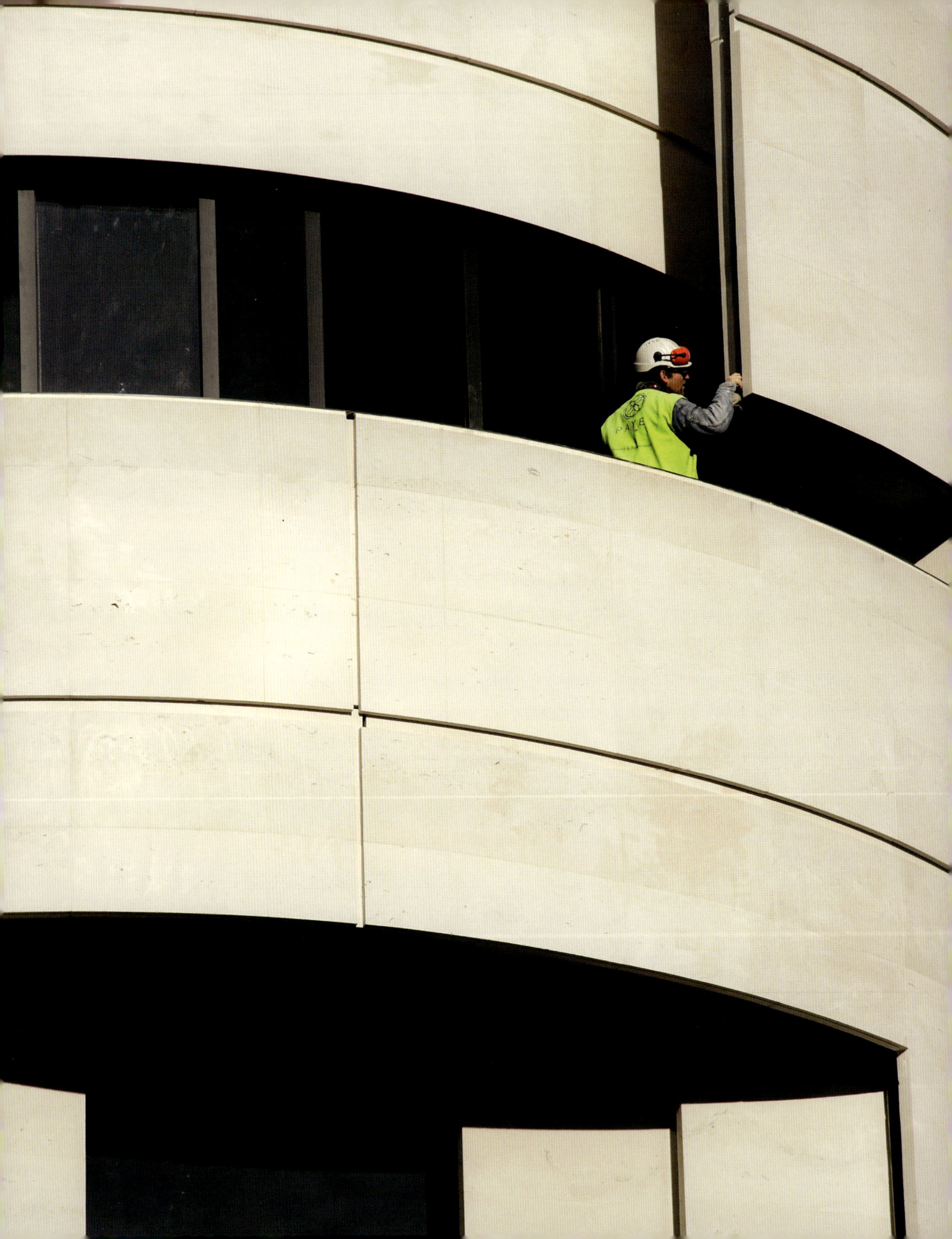
PAYE

SCAFFOLDING

The scaffolding must be designed to take into account the heavy loads that will be imposed by the hoisting, distribution, and installation of the masonry. The density of stone if wet will be in the region of 2700kg per m^3, and a pallet of stone can easily weigh in excess of 1 tonne.

A typical masonry scaffold will be five boards wide with three inner boards adjacent to the work face. The boards must not touch the façade but neither leave a gap wide enough to risk injury. The scaffold is usually fully sheeted to provide protection against dust and debris. Netting is cheaper but does not adequately contain dust created when the stone is drilled for fixings or cut. The sheets must be adequately lapped horizontally to ensure anything accidently dropped from above cannot fall through the laps and out of the scaffold.

If over a public highway, it is important to provide a double boarded bottom lift incorporating a polythene membrane to ensure that no dust, debris or water can pass through and cause nuisance below. Safety fans are also required to further protect against falling debris.

Plywood screwed into the scaffold boards is necessary to create runways for trolleys and barrows moving crates or individual stones. At least one scaffold lift needs to be free of diagonal bracing so that trolleys can be moved horizontally. Steps and ramps in the boarded runways should be avoided wherever possible.

Scaffolding must to be restrained because wind loads can be considerable. If permitted, scaffold ties are discretely installed for this purpose into the existing structure and required every 25m^2 or so, depending on conditions. On completion,

FAR LEFT
Finishing touches at BBC Broadcasting House

ABOVE
Masonry repairs in progress at Windsor Castle

LEFT
Dismantling works in progress at the V&A Museum

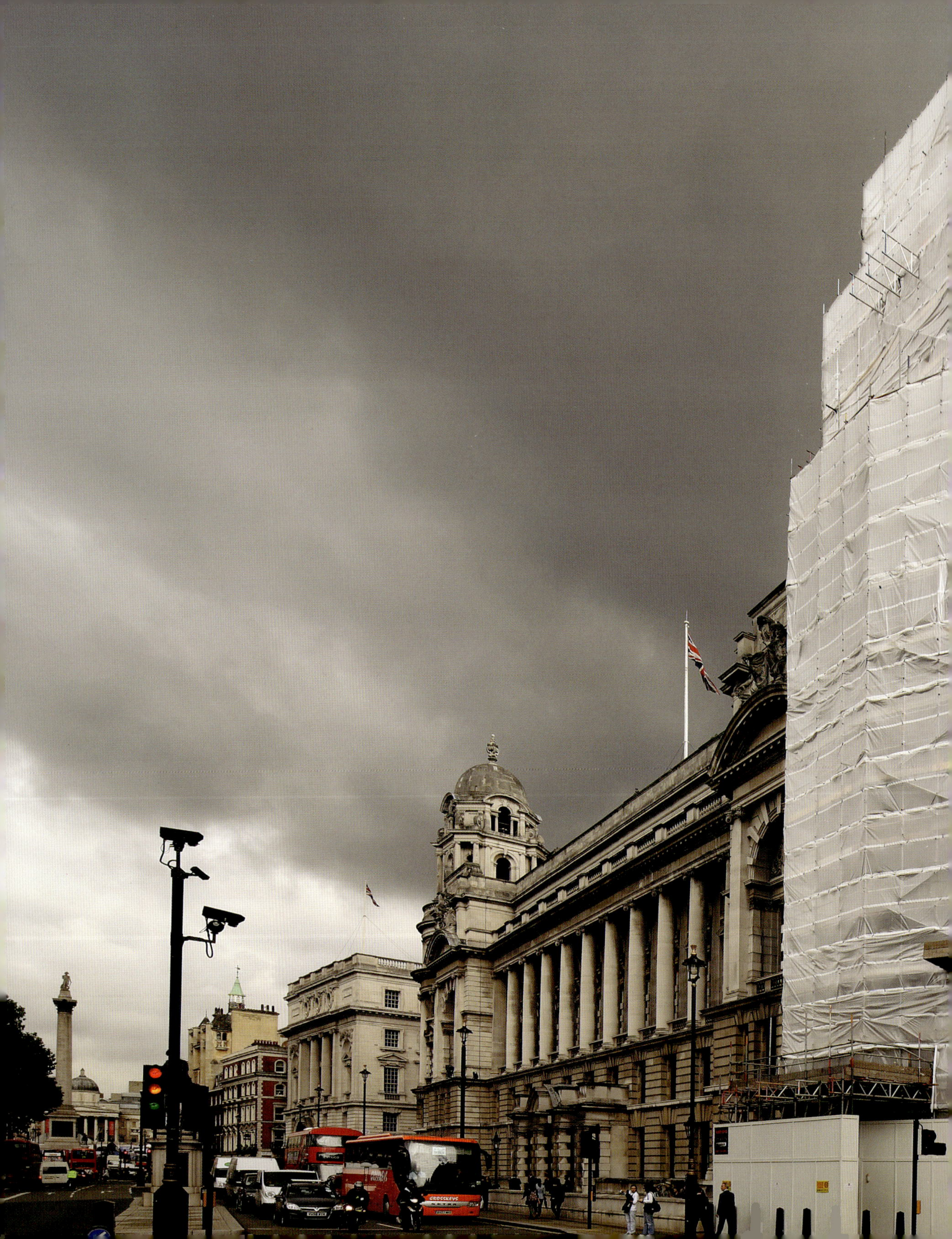

Old War Office, Whitehall

UBX
WARNING
SCAFFOLD
ALARM
0871 200 1999
www.ubxsecurity.com
PAYE
THINK SAFETY

the ties should be removed and the holes made good using an appropriate lime mortar as the scaffold is struck. Where restraint ties are not permitted, the scaffold can be designed to be freestanding without ties by erecting buttresses which are often stabilised using kentledges. Water kentledges run the risk of leaking and therefore ceasing to act as such. If used, water kentledges should be routinely checked to ensure water levels are maintained.

Restraining scaffolds to thin stone clad façades is usually problematic because the restraint fixings securing the cladding are insufficient to restrain the scaffold. Scaffold ties must therefore be fixed through the cladding into the structure behind, hopefully avoiding damage to dpc's, cavity trays, fire stopping, and cladding support/restraint fixings. In practice this is difficult to achieve without extensive investigations.

LIFTING EQUIPMENT

It is a basic principle that the stones should stay in their crates or on their pallets whilst being unloaded, hoisted and distributed from the transport delivery point to the fixing location in order to minimise the risk of damage. The other essential basic principle is that the operative must not lift any load unless he or she can do so without risk of injury. In our trade the most common injuries are those to backs and hands. In practice this means that suitable facilities must be in place before the stone is delivered to site.

Heavy stones cannot be fixed safely directly from a tower crane due to the bounce which occurs as the tower crane's cable, mast and jib deflect under load as the stone is moved. This is dangerous for the operatives and also risks damaging the stone. If a tower crane is being used it should only lift the stone and place it adjacent to its fixing location. The stone should be lifted on its pallet or crate using certified lifting straps and netted for further safety. The individual stone should then be carefully strapped and lifted by means of a block and tackle suspended from a lifting beam or running beam. Lewis pins should not be used for anything other than the final lowering of the stone into position (or if dismantling, the initial lifting of the stone from its original position) as the stone might unknowingly be fractured and fall.

FAR LEFT
Protection, access scaffolding and hoardings at Kenwood House

ABOVE
Access scaffolding in Strand, Westminster

TECHNICAL ISSUES, MANAGEMENT & ORGANISATION

TOLERANCES

The cutting of stone from large quarried blocks, referred to as primary and secondary sawing, is undertaken using large saws which can deflect when loaded during the sawing process. The deflection of the saws as they cut through the stone can be as much as a few millimetres, which is why the codes of practice and stone yards always work to an agreed tolerance which is generally now +/- 1mm. Specified tolerances of joint widths need to relate to the cutting tolerances of the stone. It will not be possible to construct a façade with consistent 3mm wide joints unless the stone is carefully worked by hand to a zero tolerance. This can be achieved, but costs will be greater than for a less demanding specification where joint widths are permitted to vary by +/- 1mm.

Frequently encountered tolerance problems include the inaccurate setting out of the grid-lines and the backing structure not being accurately constructed. It can be helpful to oversize the length of a number of ashlars on each course so that these can be cut down as necessary as the stone is being installed.

Stack bonding (where horizontal and vertical joints run continuously) has been popular even though this appears more as tiling rather than traditional masonry. It involves four stones meeting at a single point and to achieve this accurately is difficult due to the cutting tolerance. If the additional effort is not taken to minimise tolerances, the appearance will be unsightly.

An additional problem encountered at Darnley Mausoleum and elsewhere was working within the shell of an older building and having to take account of the tolerances achieved by the original 18th century masons. We found that the circular chapel was not in fact circular (by 40mm) and the horizontal floor was not actually horizontal (by 30mm). In reconstructing the fire-damaged vault we managed the tolerance issues by manufacturing and installing one course before measuring, surveying, manufacturing and installing the next. This was repeated on a course by course basis until the keystone was finally installed.

ABOVE
Emergency dismantling works at the Marshall Street Temple, Leeds

RIGHT
Lewis pins used for reconstructing the fire damaged crypt at the Darnley Mausoleum

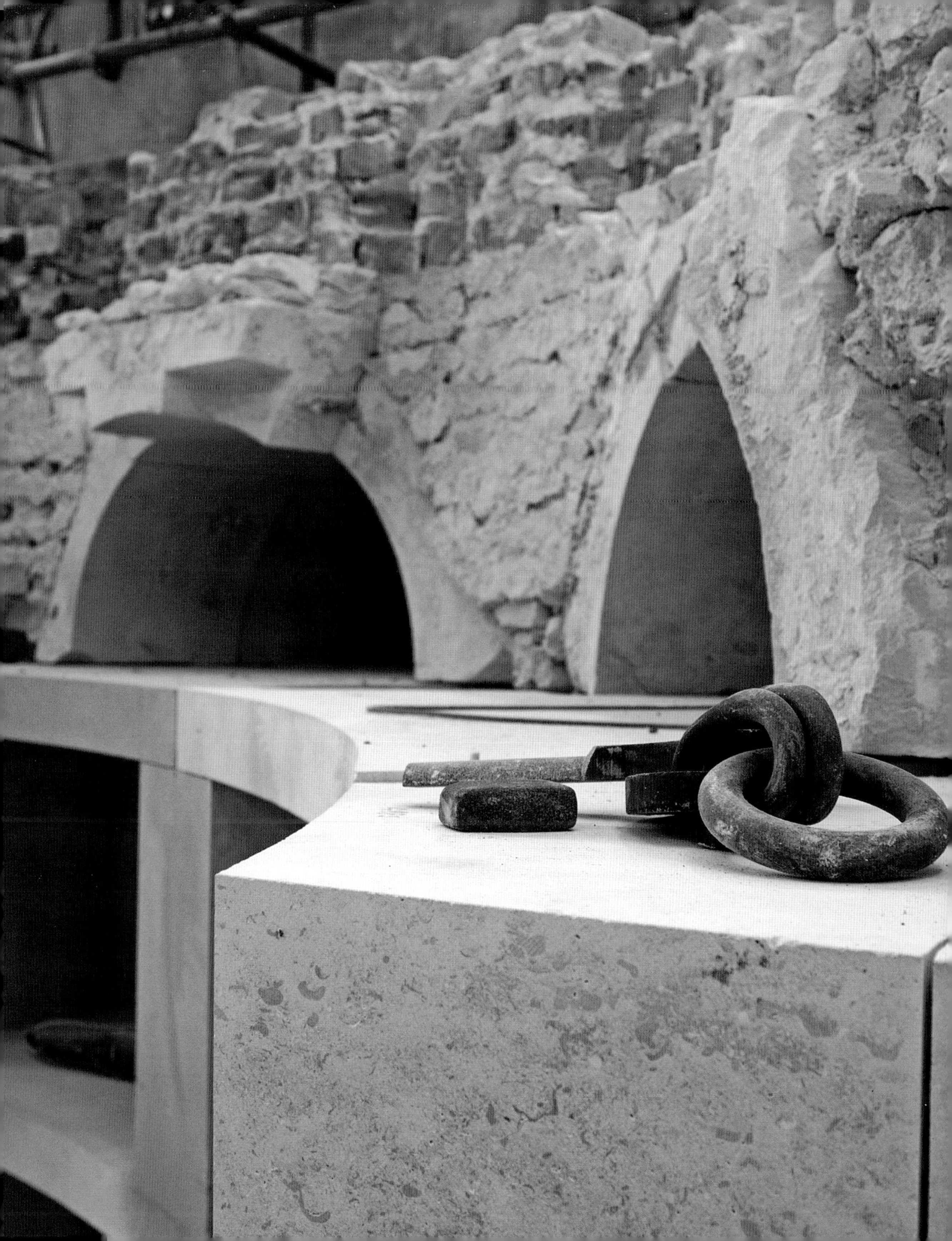

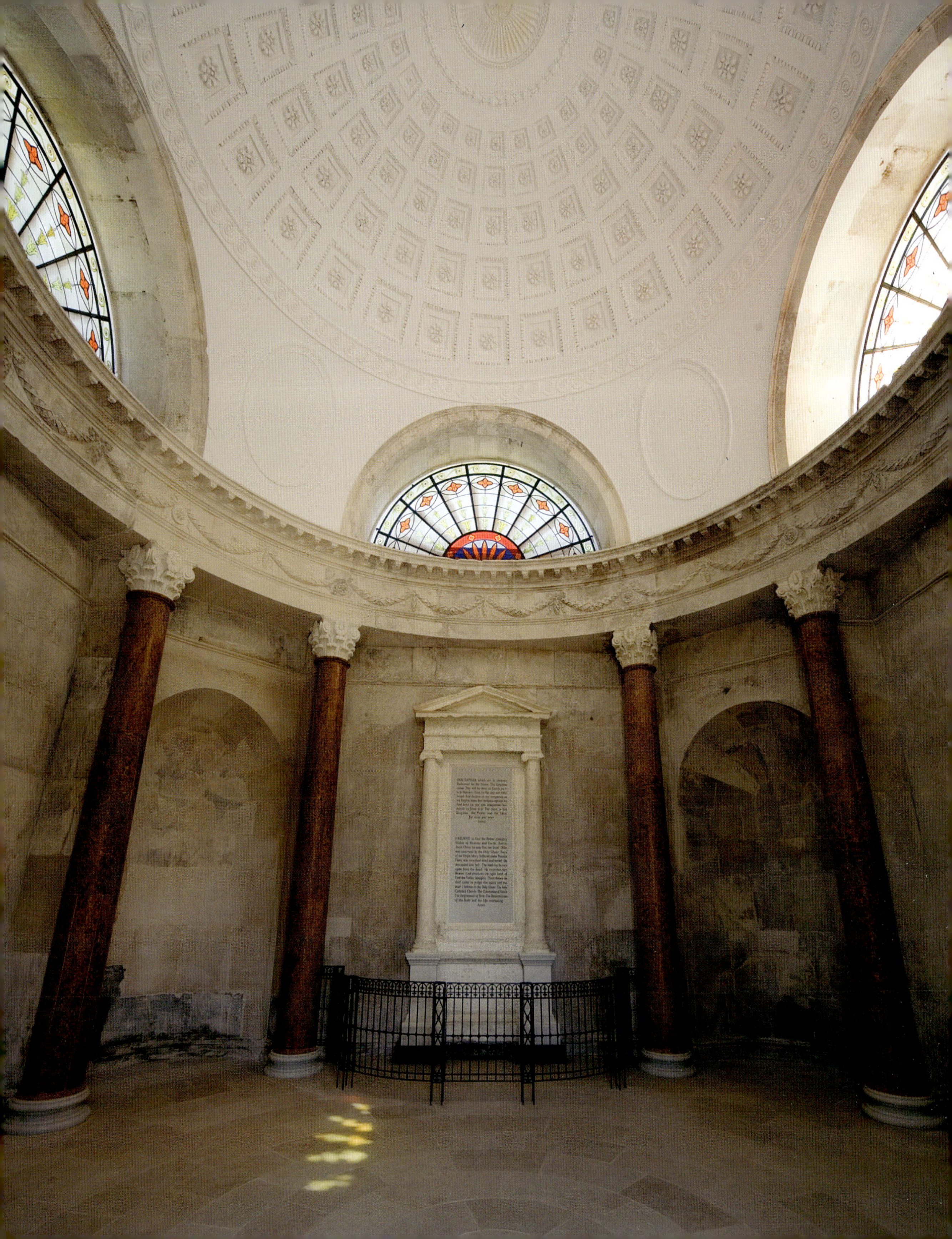

TECHNICAL ISSUES, MANAGEMENT & ORGANISATION

SHIMS

Stones are often so heavy that the mortar will be squeezed out of a joint as the stone is bedded. It is therefore necessary to space the blocks apart whilst the mortar sets, traditionally using timber shims, lead, coins (usually pennies), slates, or oyster shells. Modern shims are plastic and available in a range of thicknesses. Timber wedges are sometimes used and have the advantage that they can be later removed after the mortar has set, but we consider these best avoided because the stone arrises are too easily damaged as the wedges are extracted.

BLOCK BONDING

If the masonry forms a stone facing to a solid brick wall, the masonry will often be bonded into the brickwork every 2nd, 3rd, or 4th course. Cornice blocks usually extend to the full depth of the brick backing wall to comply with the 2/3rds supported rule. This is referred to as block-bonding. In this form of construction the brickwork needs to be built concurrently with the stonework. Usually a bricklayer will work on the inside face of the wall as the masonry is installed from the outside face.

GROUTING

Good practice dictates that all masonry joints are fully filled as work proceeds. It is easy for stones to be laid on a full bed of mortar, however the perpendicular vertical joints are less easy to fill so often need to be grouted. This is achieved by pointing the vertical joint and pouring in behind a diluted lime putty, stone dust and sand liquid grout to fill the void. It has always been too common for the grouting of vertical joints to be omitted (by modern and medieval masons alike) with the result that the pointing of the vertical joints soon fails.

Grouting is also used to fill voids in rubble cored walls, etc.

MOVEMENT JOINTS IN TRADITIONAL MASONRY

Traditional block-bonded brick-backed masonry walls are so thick they do not heat and cool enough in a 24 hour cycle to stress the masonry. This is why so many large traditional façades have survived for hundreds of years without thermal cracking. It is incorrect to attribute the lack of thermal cracks to the use of lime mortars.

LEFT
Reconstructed crypt vault of the award winning Darnley Mausoleum

ABOVE
Reconstruction works within the fire damaged chapel at the Darnley Mausoleum

TECHNICAL ISSUES, MANAGEMENT & ORGANISATION

THIN CLADDING SYSTEMS

Until recently the codes of practice published tables stating the required thickness of cladding for various types of stones in specific circumstances. Typical minimum cladding thicknesses above a height of 3.7m were 75mm for limestone and 40mm for granite. Only in exceptional circumstances could these thicknesses be reduced. However, it is now permissible to reduce these thicknesses providing the use of thinner sections can be verified by technical data and calculation.

The codes of practice until recently stated that masonry cladding must be supported every storey to avoid Ronan Point type progressive collapse in the event that masonry was displaced causing the entire cladding system above to fail. This consideration seems to have been lost in the latest codes of practice. It was permissible for up to three storeys to be supported at one level.

Loads are generally transferred to the framed backing structure into floor beams or floor slabs.

Horizontal movement joints are usually provided below the cladding support fixing positions. Vertical movement joints are typically required every six to nine metres around the perimeter of the façade, and close to external returns. Movement joints are usually 10-15mm wide and pointed with a suitable sealant coloured to match the stone. Depending on the type of sealant used it is often necessary to prime the stone surfaces to prevent the masonry becoming stained. It is sometimes possible to finish the mastic sealant with matching stone dust in order to disguise it as a normal masonry joint (though this is rarely fully successful). In a façade intended to resemble an historic building, movement joints can appear unsightly. It is often best to conceal horizontal joints in shadow lines under cornices where they won't be noticed. Alternatively, they can be installed above a cornice if there is concern that its position below would detract from the general

ABOVE
Window detail at BBC Broadcasting House

RIGHT
Completed radiused ashlar Portland stone façade at BBC Broadcasting House

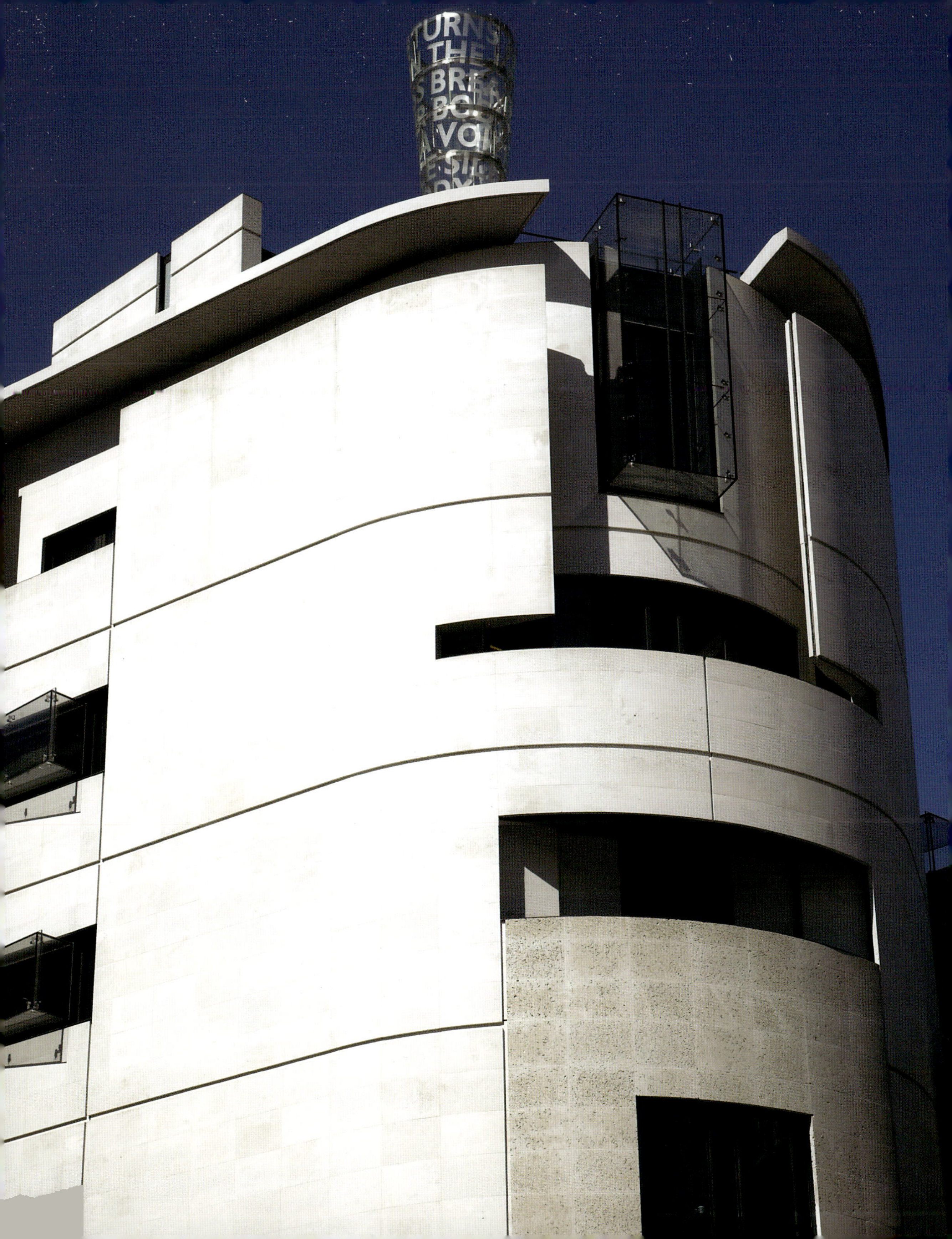
URNS
THE
S BREA
A VOI

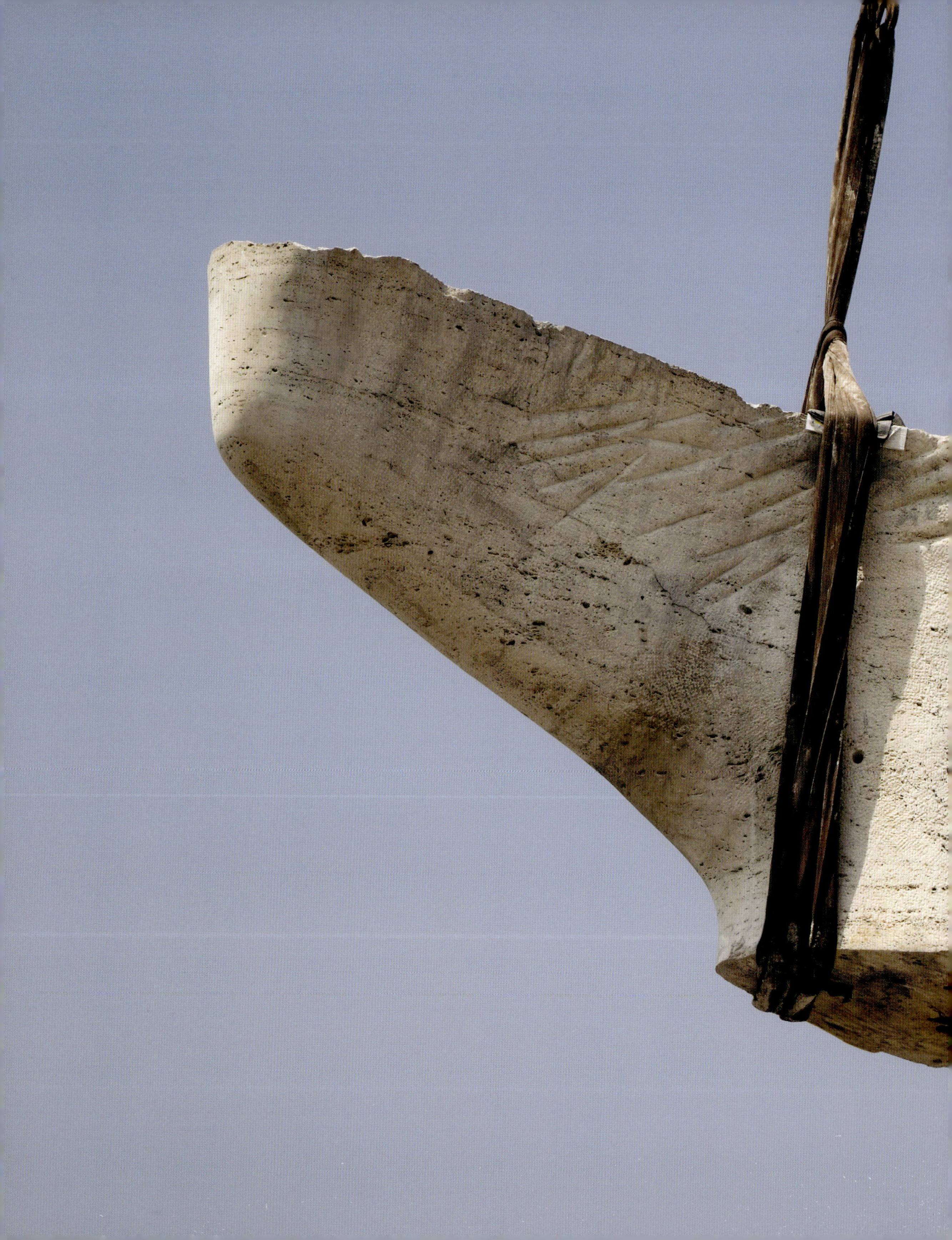

Reconstruction of the Travertine Henry Moore Arch, Hyde Park

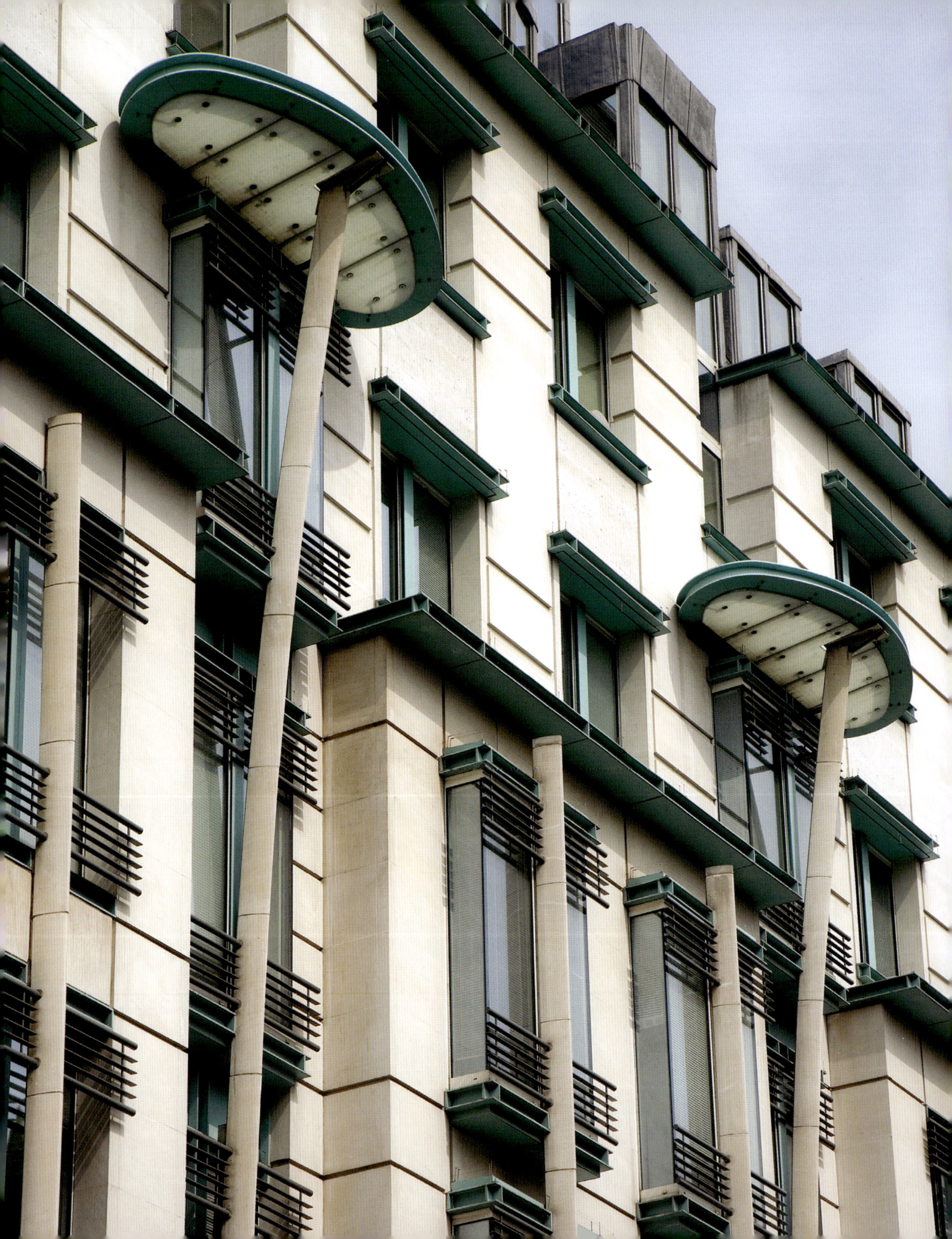

TECHNICAL ISSUES, MANAGEMENT & ORGANISATION

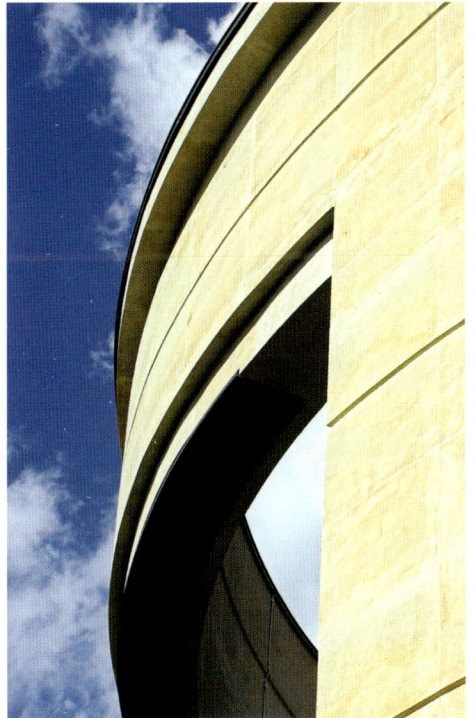

appearance. Vertical movement joints are also unsightly and often located behind convenient rainwater pipes or hidden in returns. Some designers require movement joints to be omitted entirely from thin cladding systems provided that technical data and calculations substantiate that this can be safely achieved.

All individual cladding units must be restrained at least four times. In practice this is achieved using stainless steel dowels and cramps positioned within the masonry joints.

FAR LEFT
Completed Portland stone ashlar façade in Pall Mall, Westminster

LEFT AND ABOVE
Bath stone cladding in contemporary design to complete the unfinished All Saints Church, Dulwich

115

TECHNICAL ISSUES, MANAGEMENT & ORGANISATION

CASTING CONCRETE AGAINST MASONRY

Casting concrete against masonry is to be avoided because the stone will absorb moisture and other contaminants from the concrete. Moisture will move through the masonry to evaporate from its exposed surface, and this will not only stain the surface but also cause salts to crystallise at the point where the moisture evaporates. This is generally about a millimetre or so beneath the surface. Salt crystals will often grow larger than the natural pores within the stone, causing the stone surface to spall. This is known as crypto-florescence.

Casting concrete exerts considerable lateral loads which have the potential to severely damage historic masonry. If concrete must be poured against historic masonry it must be protected with a suitable membrane and be stabilised against lateral forces.

Limestones contain residual soluble organic material originating from the sea creatures from which the stone was formed. Moisture moving through the masonry to evaporate from the stone surface will draw these organic residues to the surface creating an unsightly stain. Ultra-violet sunlight will break down these residues and cause the stain to dissipate over a period of months. If required, these stains can be removed using a poultice.

BEDDING CILLS

Loads being transferred through a stone need to bear as uniformly as possible, otherwise the stone will fracture. Fully bedded stooled cills therefore crack where only the ends are loaded. One way to mitigate this differential loading is to solidly

RIGHT
*Portland stone detail,
St Paul's Church,
Mill Hill*

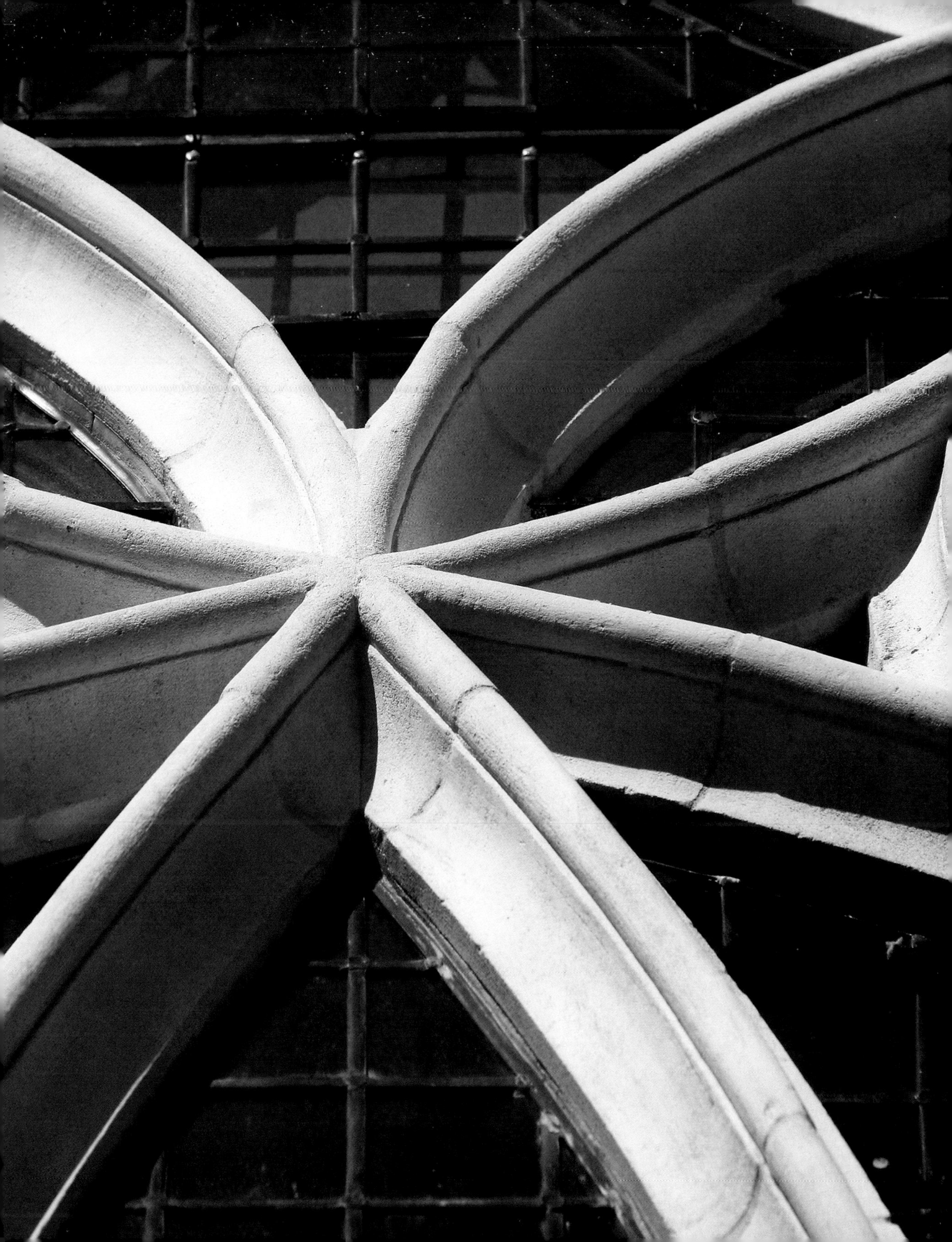

TECHNICAL ISSUES, MANAGEMENT & ORGANISATION

bed the stooled ends only and leave the bed open under the window. Even though the open bed is often surface pointed, the void behind is often enough to ensure the stooled end can take up the load and not fracture against the unloaded centre section.

CANTILEVERED STAIRS

Though they are designed to appear so, cantilevered stairs are not cantilevered. Each step bears on to the step below and onto the landings which transfer the loads into the surrounding walls. Whereas the embedding wall provides torsional resistance to keep each step in position, the staircase basically acts in compression. Landing slabs are usually supported on two sides (sometimes three if the span is small or the stone is unusually large) and joggle-jointed together. Occasionally a landing slab will act as a true cantilever where imposed loads either do not exist or are relatively small.

Cantilevered staircases are delicate structures which are easily fractured should any differential movement occur within the embedding staircase walls. Another cause of failure is where the landing slabs fail and no longer support the steps. Occasionally the grout between steps can fail and compromise the transfer of loads and support, fracturing a step or two.

Cantilevered staircases can be repaired but a thorough understanding of the cause of failure, the imposed loads, and the methods of repair are needed to ensure success.

Metal balustrades are often embedded into the steps. If corrosion occurs, the ends of the steps will crack and spall.

Right
Radiused cantilevered staircase and concave landing detail in Moleanos limestone

These defects can usually be indent repaired, treating the corroding metalwork. If the stone staircase displaces, the metal balustrade often helps to take up additional load and thus the two components work together as a composite structure. When this has occurred, the metal balustrade will often resonate if gently struck.

If it is necessary to replace a stone step it will also be necessary to cut the metal balustrading. The embedded lug should be retained, set into the replacement step, and welded to the balustrading upon installation of the new step. To effect a good quality repair, the weld should be fully ground down to become invisible.

Wear can sometimes be excessive and it is not uncommon to find that the stone steps have been previously indent repaired. Occasionally we find that these indents have been cut too deeply into the step, causing it to fail.

It is sometimes possible to stitch repair an individual fractured step, but if the fracture extends across two or more steps it is usually considered appropriate to replace the steps.

Steps can be carefully cut out and replaced after the staircase has been braced and supported. The design of a suitable temporary support and bracing system needs to be carefully considered because steel scaffolding will deflect greater under load than the stone staircase will be able to without further fracturing. All temporary loads must therefore be transferred to the surrounding walls without bearing onto the staircase below.

Landing slabs can be difficult to replace due to their substantial size. Safely manhandling such large units through existing structures without the use of cranes is at best problematic and often impossible. For this reason it is sometimes necessary to replace landing slabs with insitu cast concrete appropriately finished in matching natural stone.

Cantilevered staircases were often constructed in Portland stone. Occasionally York stone would be used, particularly for servants' staircases. 'Above stairs' they were often painted and their soffits lined with paper, presumably in attempt to raise their appearance to that of marble.

We are often asked to remove the paper and paint, and to subsequently clean the stone.

However it is always necessary to manage the clients' expectations because an old Portland stone staircase will never resemble a shiny polished marble. Marble has historically been used to construct cantilevered staircases in many important historic houses. Other imported stones have been used to construct cantilevered staircases in recent years.

In many occupied buildings, the cantilevered staircase forms part of a fire escape route. As the temporary support works will generally obstruct access, it is often necessary to ensure alternative means of escape are available whilst works are being undertaken.

TOP
Cantilevered staircase and bronze balustrade detail

BOTTOM
Award winning York stone cantilevered staircase with protection and temporary support at The National Gallery

Cantilevered steps set into brickwork at Luxford House, Sussex

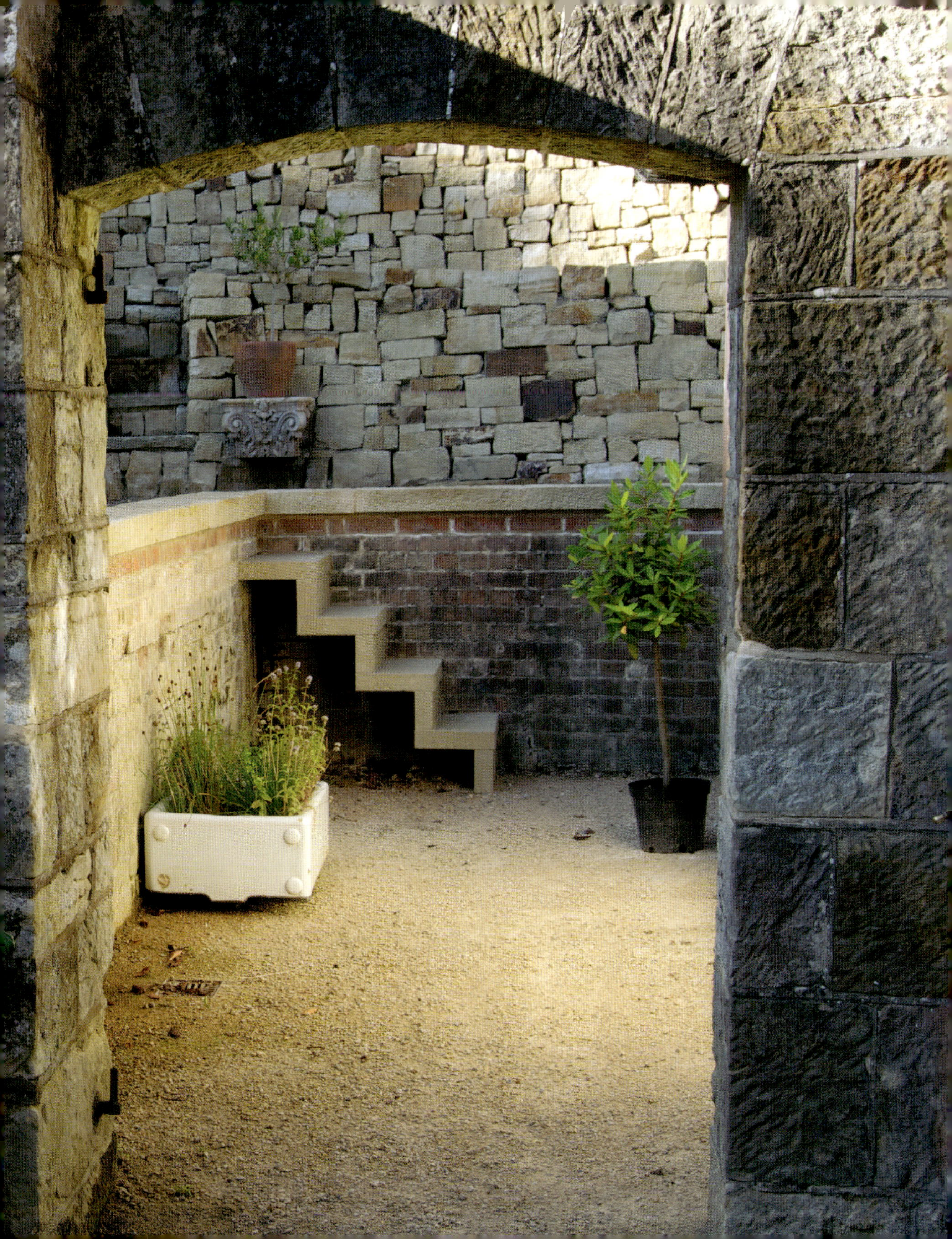

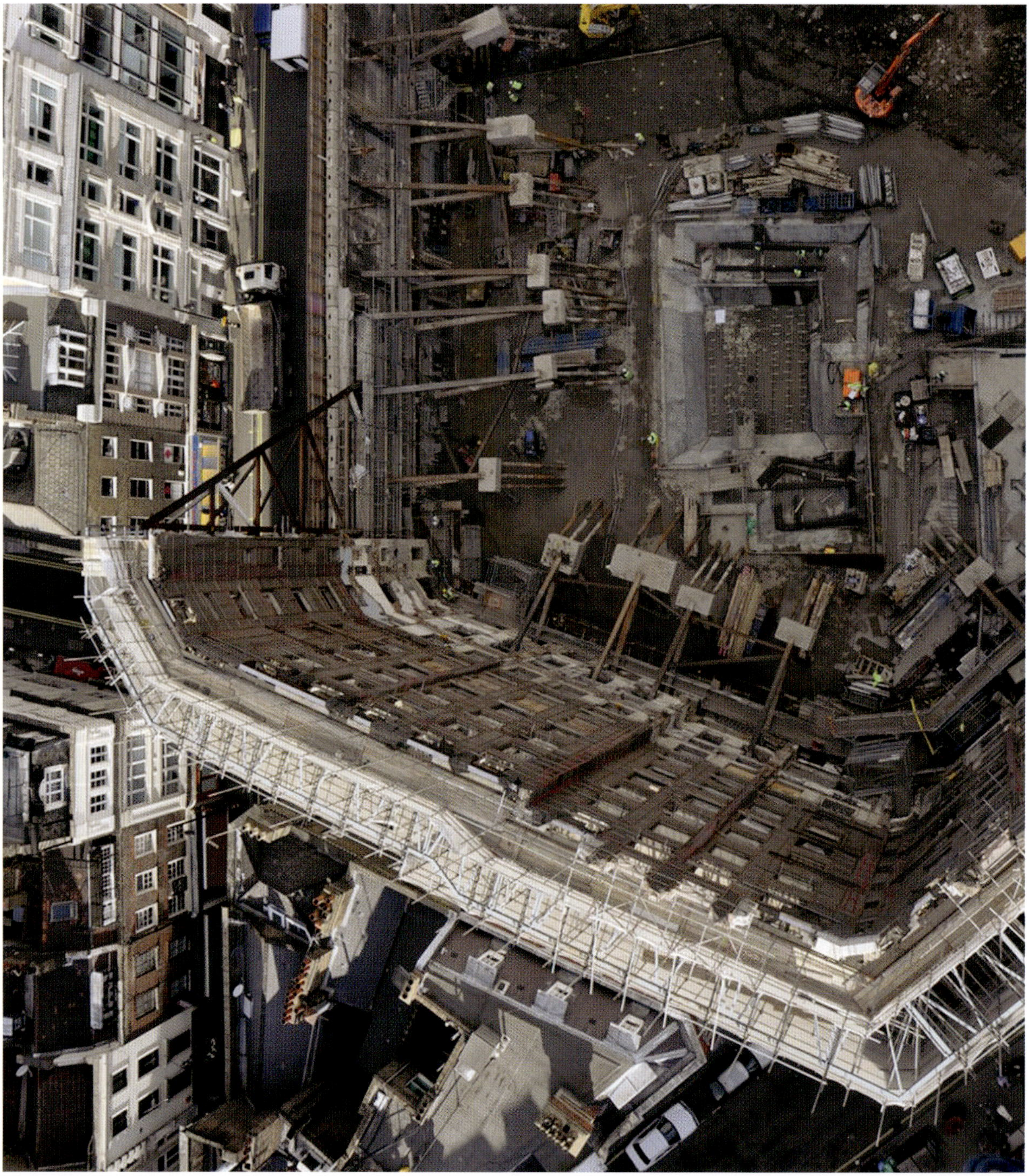

TECHNICAL ISSUES, MANAGEMENT & ORGANISATION

PROTECTING A RETAINED FAÇADE AFTER DEMOLITION

When the building behind a retained façade is demolished the back face of the façade becomes exposed and susceptible to damage from exposure to rain. Embedded steel will potentially corrode causing local damage of the masonry. Rainwater saturating the façade might also cause efflorescence which, if left for months, will be difficult to remove. Soluble organic residues naturally occurring within the stone will be drawn to the surface causing staining if rainwater is allowed to become absorbed by the retained façade.

It is therefore necessary to protect the top and rear face of the retained façade with roofing felt secured by means of timber battens whilst exposed to the elements.

The management of rainwater is also important during the later stages of construction of the new structure. It is important to keep the retained façade dry to avoid staining. On poorly managed sites, the roof structure is often completed before the rainwater pipes are installed, resulting in the saturation and staining of the façade. The reason for this is probably connected to the principal contractor's procurement strategy and the interfacing of many different specialist subcontractors. These problems can be avoided by the use of temporary rainwater pipes and adequate site management.

Rainwater should be prevented from ponding on the new concrete floor slabs and saturating the retained façade during the construction of the new structure. The use of bunds can be useful to ensure this does not occur.

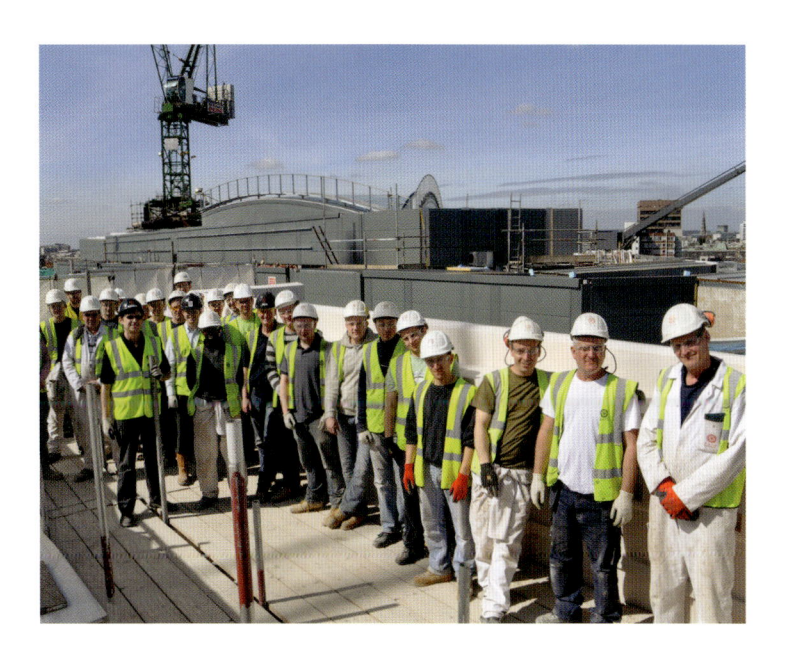

MANAGING DIFFERENTIAL MOVEMENT BETWEEN A RETAINED FAÇADE AND NEW STRUCTURE

A façade relieved of its loads by the demolition of the building behind will have a tendency to rise as the previously loaded bearing strata heaves as the loads are reduced. This creates the potential for differential movement. Furthermore, the new structure behind the façade will often bear on its own foundations, providing only restraint for the retained façade. The new building will also settle onto its new foundations during construction and thereafter, requiring adequate provision for differential movement between the façade and the structure behind. The design of the restraint fixings between the two structures must therefore provide for all differential movement.

The new roof structure will usually be supported off the new structure and

LEFT
View from the tower crane of the retained façades, Regents Palace Hotel, Regent Street

ABOVE
Masons taking a break from replacing nine thousand faience blocks, Regents Palace Hotel, Regent Street

TECHNICAL ISSUES, MANAGEMENT & ORGANISATION

TECHNICAL ISSUES, MANAGEMENT & ORGANISATION

FAR LEFT
Retained faience façade at the Regents Palace Hotel showing retention system and weather protection

BELOW
Retained faience façade at the Regents Palace Hotel showing weather protection marked up to identify positions of embedded steel frame

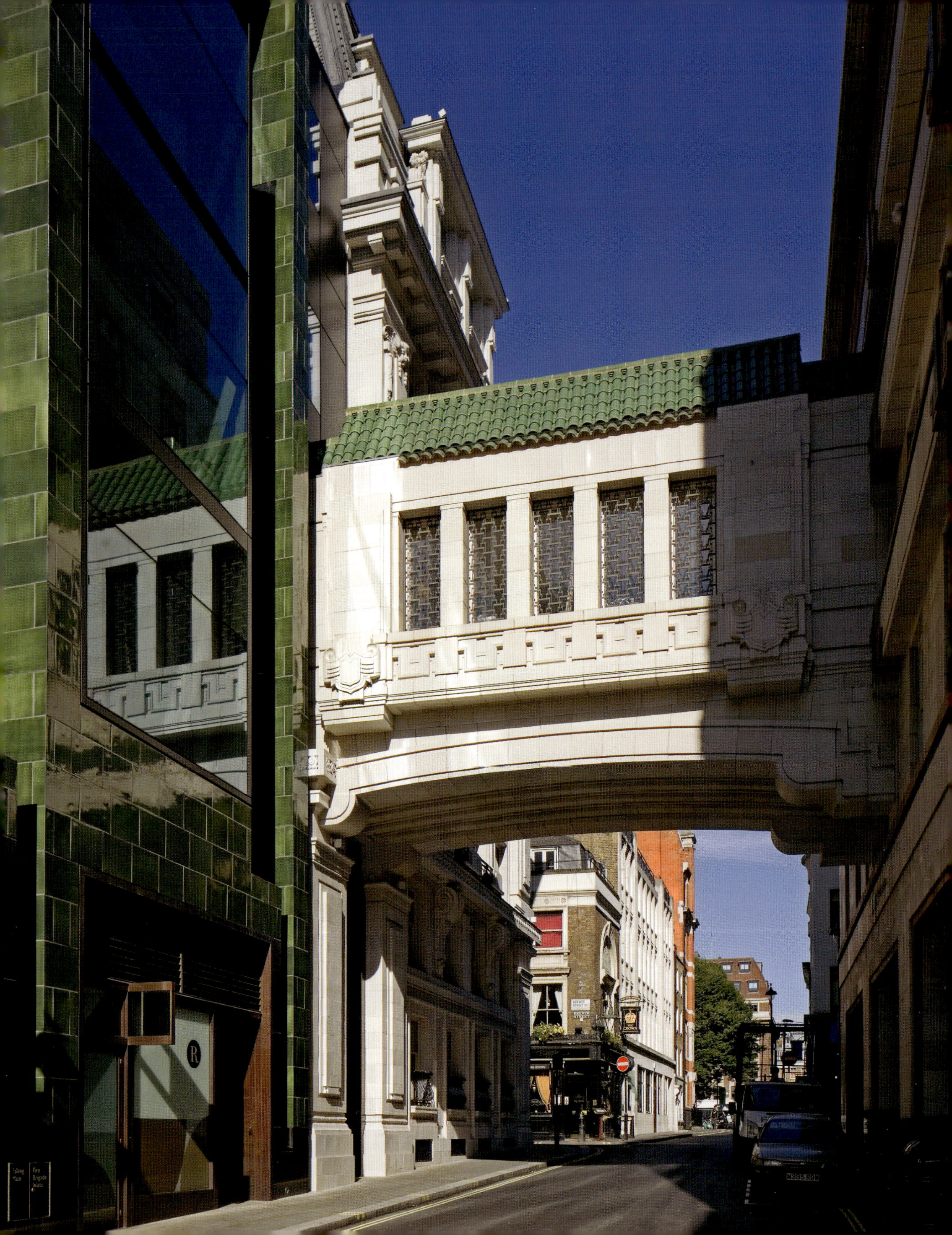
R

will therefore need to be designed to be independent of the retained façade. Weathering details will require careful consideration in gutters behind the retained façade designed to accommodate the differential movement.

DISMANTLING, STORING AND RECONSTRUCTING HISTORIC FAÇADES

Whereas the incorporation of a retained façade onto a new building is common practice, PAYE are being increasingly employed to dismantle, store, and reconstruct historic façades. This is because it is often more efficient to temporarily remove the façade in order to provide better access for other works. During reconstruction the façade is often discretely adapted, for instance to raise the height of shopfronts or upper storeys.

Local planning authorities are understandably nervous when asked to approve such schemes as there have been cases where the dismantling has been undertaken by unskilled inexperienced labour without due care nor supervision, a famous example being where the masonry from the important Grants of Croydon façade was dumped in a loose pile without protection, care, nor the retention of any records. The stone was unsurprisingly found to be too damaged to be used in the reconstruction of the façade. Many large stones were fractured and all the stones were badly chipped. Sadly, the historic masonry was never reused and the façade was eventually reconstructed with new stone.

FAR LEFT
Repaired faience bridge with replaced faience profiled roof tiles, Regents Palace Hotel

ABOVE
Dismantling the Aston Webb screen wall at the V&A Museum

LEFT
Dismantled masonry at the Marshall Street Temple, Leeds, awaiting reconstruction

RIGHT
Travertine Henry Moore Arch, Hyde Park

With correct procedures and a skilled workforce it is possible to successfully dismantle, store, and reconstruct historic masonry. Numerous such projects have been successfully completed by PAYE including the Henry Moore Arch, Crown Estate façades in Piccadilly and Regent Street, the Aston Webb screen wall at the V&A Museum, and various façades in the City of London. PAYE often support listed building consents and planning applications by preparing method statements demonstrating the skills of our workforce and that the work can be safely undertaken.

CORRECT PROCEDURES INVOLVE:

- Accurate surveys, identifying all joints and allocating reference numbers to individual stones. The general convention is that the masonry courses are numbered from the lowest and individual stones are numbered left to right.

- Indelibly marking the unique reference numbers on a hidden face whilst the masonry is dismantled.

- Recording details of any stone that needs to be repaired or replaced.

- Carefully packing the stones onto pallets giving each a reference number, and recording a schedule identifying which pallets contain which stones. To protect the arrises of all the stones, non-compressible packing strips should be provided so that no stone bears on or rubs against another. This also provides an air gap around each stone which helps to avoid the formation of algae on the masonry surfaces. The pallets should be shrink-wrapped in a manner which will protect the stone in transport from saturation from rain but will still allow air to circulate. Lifting and distribution equipment is necessary, and runways will need to be ply-boarded.

- Sufficient record photographs.

- Loading pallets from the storage gantry onto flat-bed trucks for transportation to the storage facility where they will be placed in a manner that will permit them to be returned to site in reverse order.

- Adequate records must be maintained.

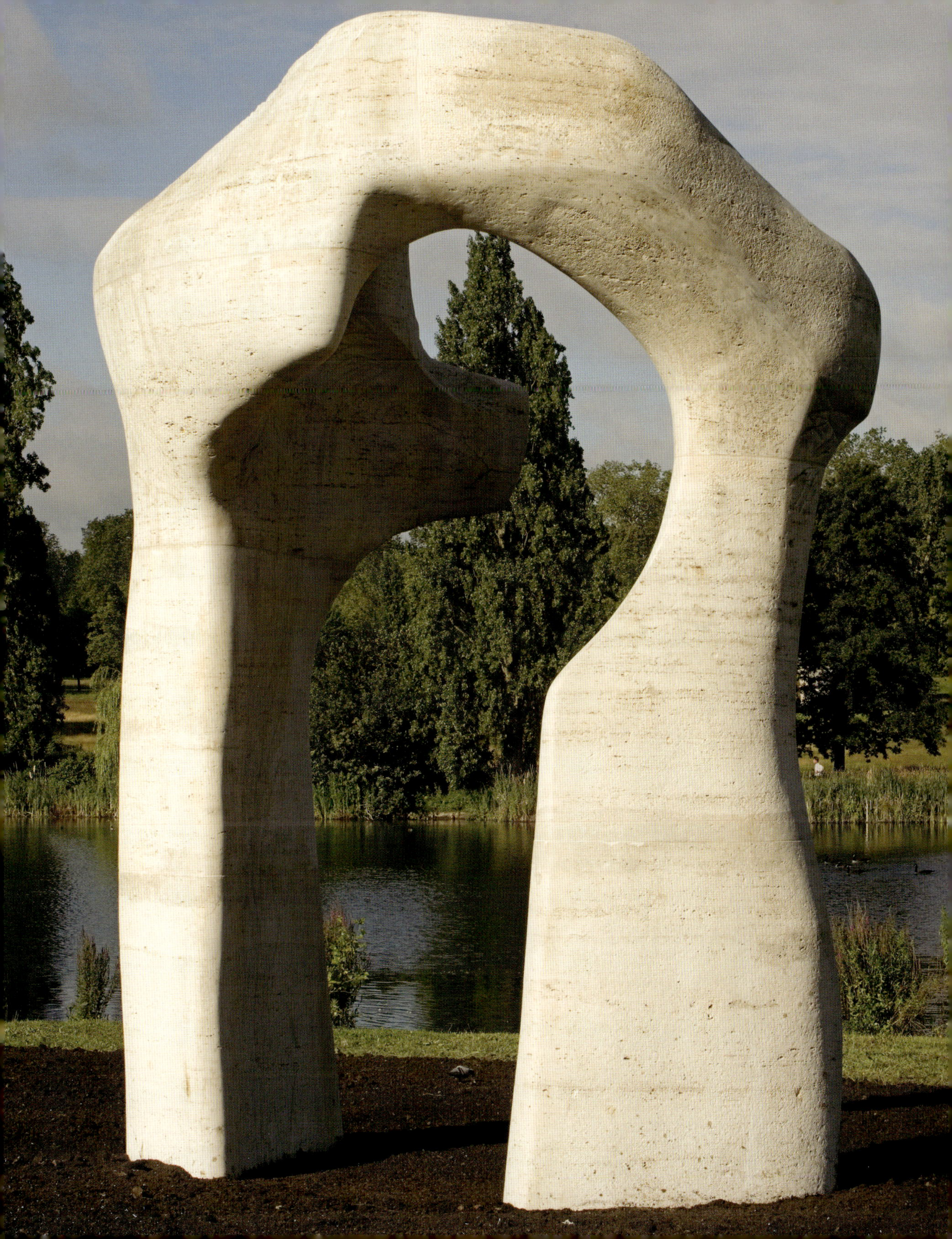

Proposed South Elevation
1:35

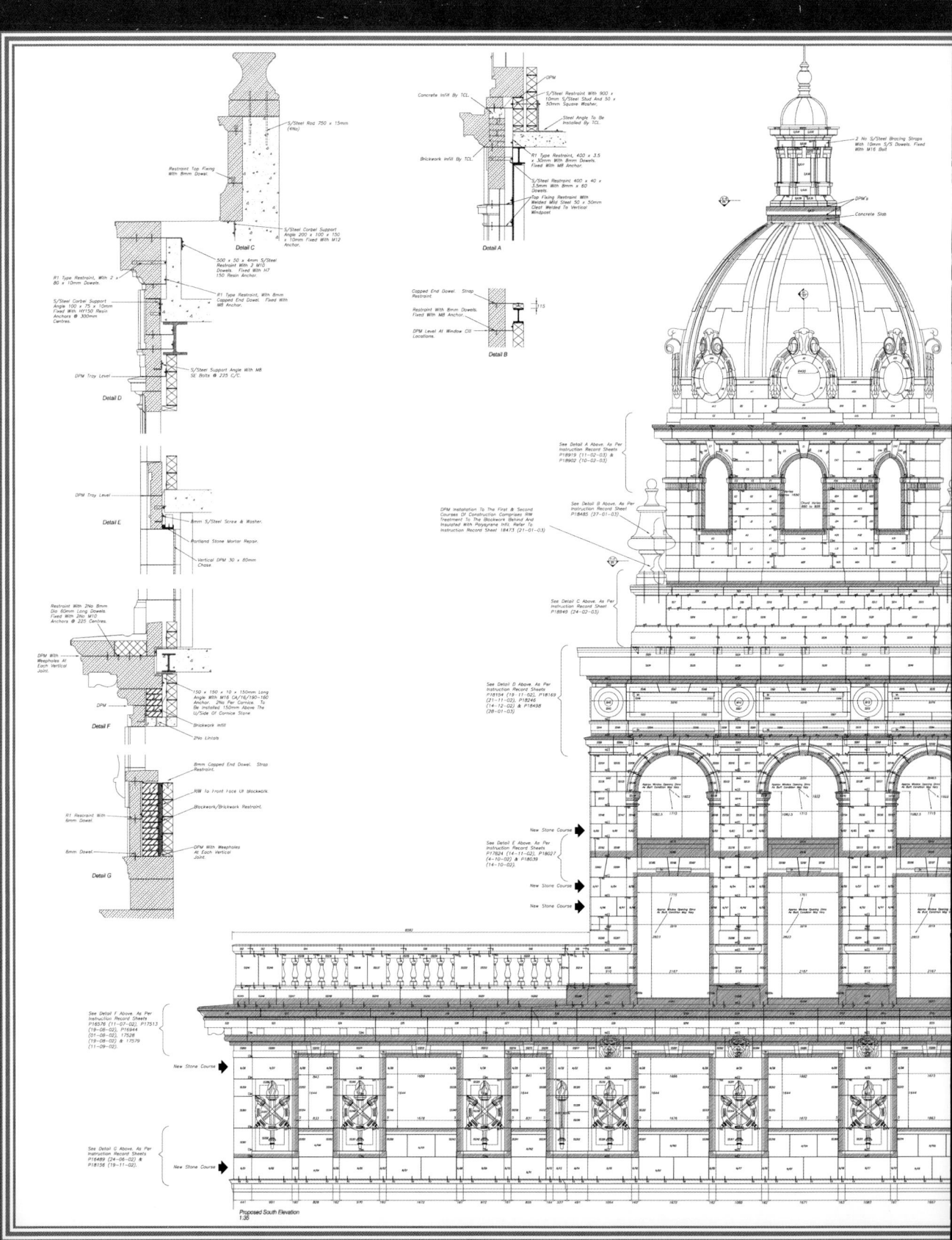

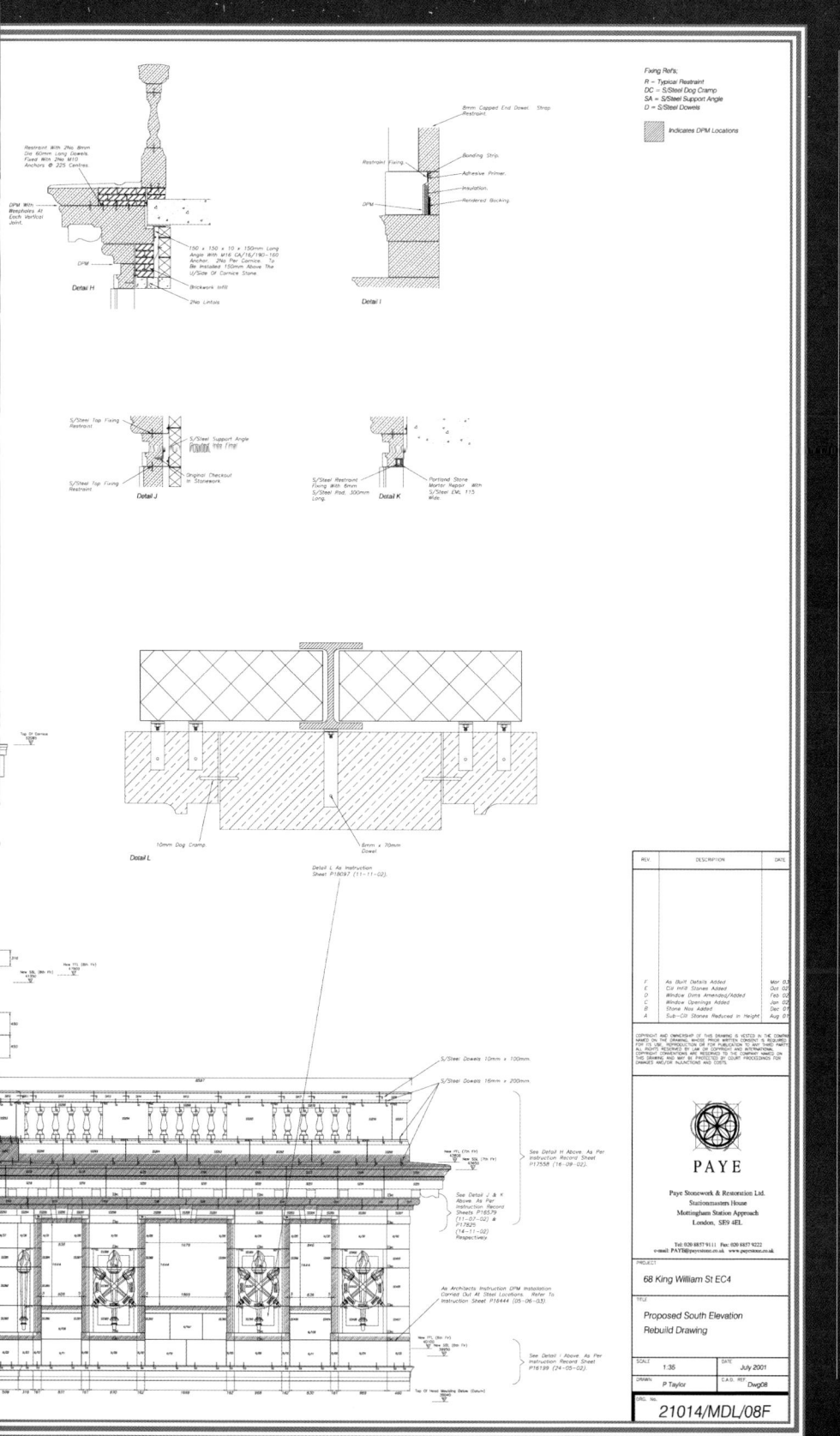

Survey drawing showing proposals for dismantling, reconstructing and extending an existing Portland stone façade in the City of London

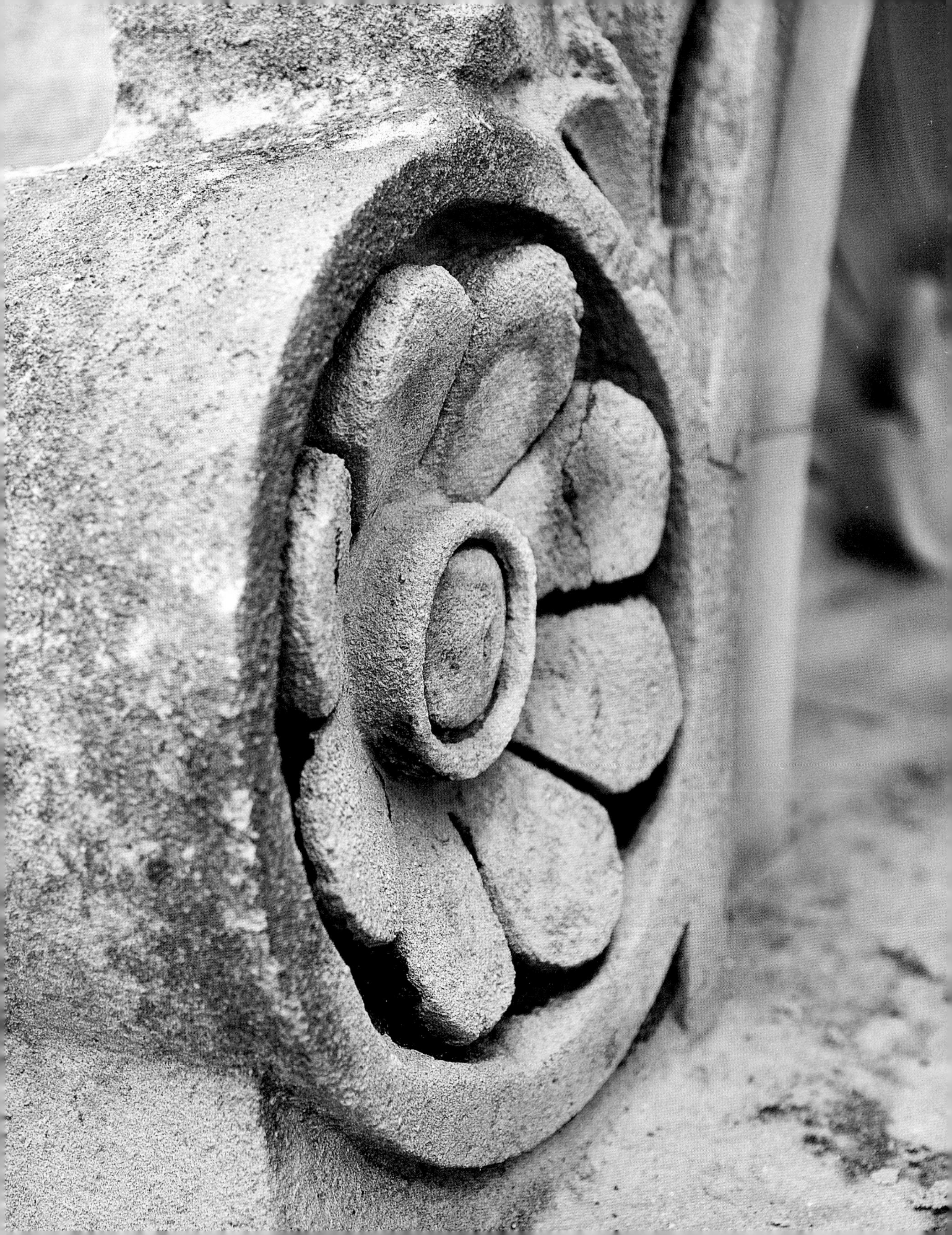

TECHNICAL ISSUES, MANAGEMENT & ORGANISATION

HISTORIC USE OF METALS

Wrought iron has a carbon content of less than 1%, is resistant to corrosion and strong in tension. If broken it will reveal a fibrous structure. Cast iron dates from the late 18th century and is smelted at much higher temperatures than wrought iron. Its higher carbon content of up to 5% makes it rigid in compression, but brittle and weak in tension. Broken cast iron looks crystalline. Mild steel has been used from the late 19th century and is the most common form of steel used in construction. Its cost is relatively low. The carbon content of mild steel is between 0.16% and 0.29%. The higher the carbon content, the harder, stronger, more brittle and more difficult it is to weld. Carbon steels corrode when exposed to moisture and oxygen. Stainless steel is an alloy with approximately 10% or more chromium and does not corrode as easily as mild steel. Under most current codes of practice, stainless steel should be used for any fixings which are embedded into an external masonry façade.

FAR LEFT
Floral detail in Portland stone

ABOVE
Refurbished wrought iron railings

LEFT
Portland stone façades at Horseguards Parade and beyond in Whitehall

133

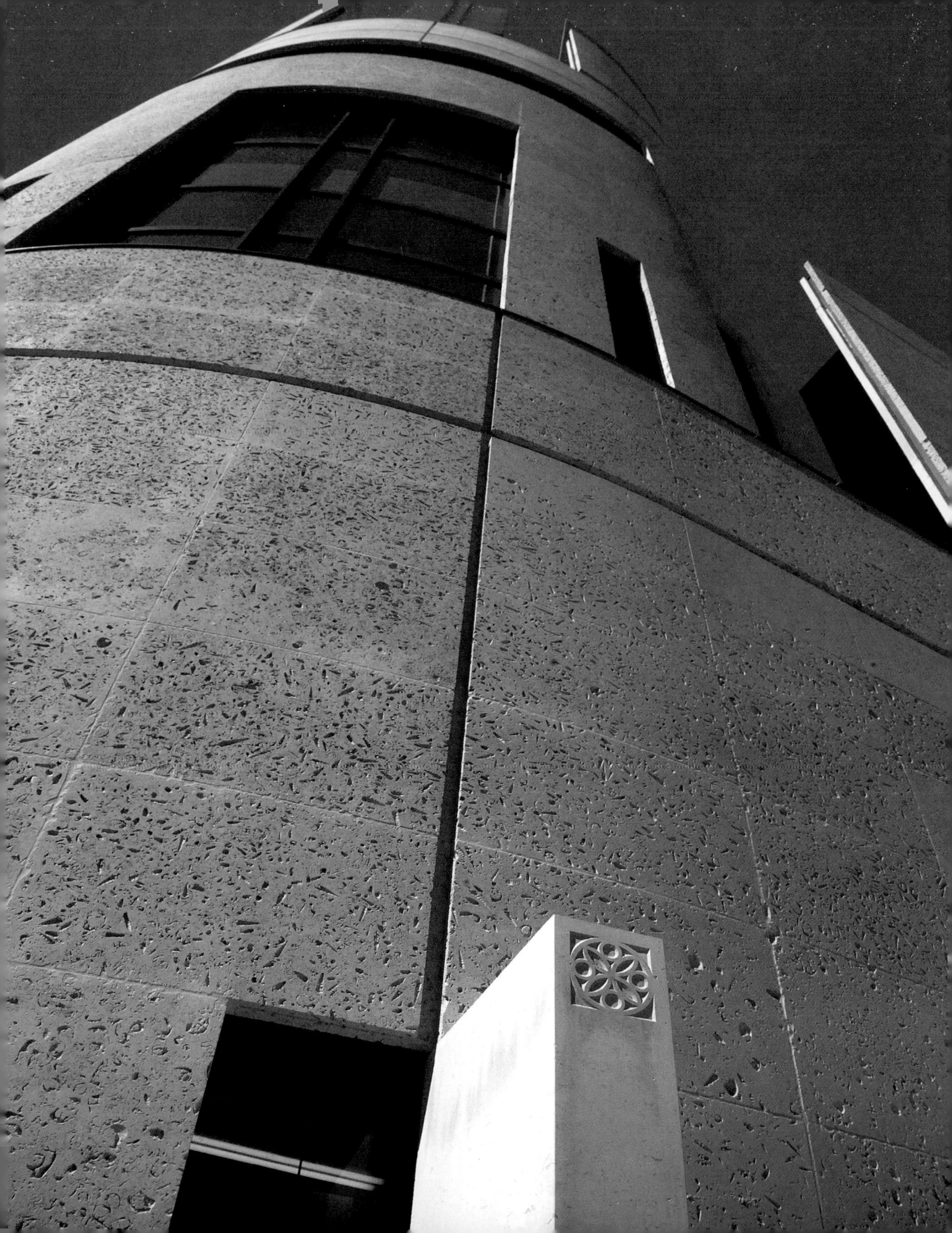

TECHNICAL ISSUES, MANAGEMENT & ORGANISATION

CORRODING STEELWORK

From the late 19th century and into the early 20th century, steel beams and columns were built into loadbearing façades. These were not fully framed structures. Examples of such façades repaired by PAYE include the Hippodrome and the UK Supreme Court. Fully framed structures were first seen in London early in the 20th century.

The Ritz Hotel in Piccadilly was constructed in 1906 and is a famous example of an early fully framed façade. It was over-designed to be a load-bearing façade and to have a fully framed steel structure embedded within it. The unnecessary cost and weight was recognized, and thereafter, façades with embedded fully framed structures gradually became ever thinner and lighter with greater areas of glass. Examples of such buildings are numerous and include BBC Broadcasting House, the Lutyens designed Reuters Building at 37 Fleet Street, and numerous other façades in Regent Street.

Regent Street Disease is a term used to describe the malady of corroding embedded steel frames causing fractures to appear in masonry façades. The problem was encountered in Regent Street during the 1930's, a few decades after the construction of these façades. PAYE have repaired hundreds of these façades in Regent Street and elsewhere. With improved rainwater management and properly implemented maintenance strategies these façades now perform safely and relatively economically.

Rust forming on the surface of a corroding metal can increase by up to thirteen times the volume of the original metal. If the metalwork is tightly grouted into the masonry, the corrosion will stress

FAR LEFT
Roach bed Portland stone featuring masons mark at BBC Broadcasting House

ABOVE AND LEFT
Ritz Hotel, Piccadilly

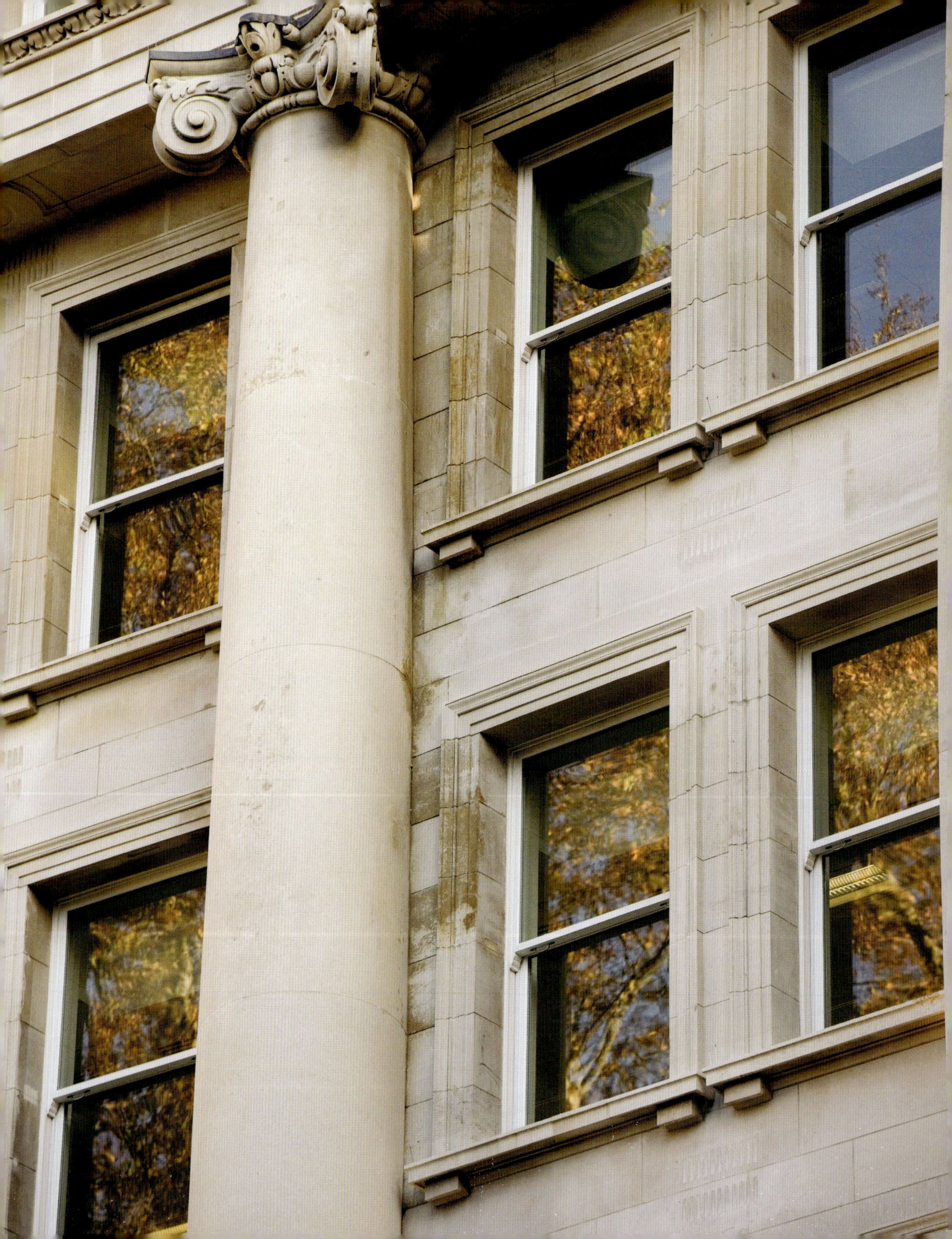

TECHNICAL ISSUES, MANAGEMENT & ORGANISATION

the stone causing it to fracture. Further corrosion will displace the masonry, potentially causing it to spall and risk sections falling to ground level. As fractures widen, more rainwater and moisture can penetrate leading to an increase in the rate of corrosion.

The amount of corrosion forming on the surface of a corroding steel component will usually fracture and displace the masonry by a corresponding amount. Though not particularly accurate, an initial assessment can therefore be made not just of the amount of surface corrosion but also the loss of section profile. This can be verified at the appropriate stage by carefully cutting out a section of fractured masonry and exposing the corroding steel.

It is rare to find that the steel has corroded to the extent that it must be replaced, though at PAYE we have known this to be necessary. Such façades were either in exposed marine environments or particularly poorly detailed and constructed. It is more common to find that the steel only requires preparation and treating, though occasionally we find it has corroded to an extent where it needs to be repaired by welding stiffening plates. The criteria which has been used as a rule of thumb by structural engineers to decide whether it is necessary to weld plates to strengthen corroding steel is where more than 15% of the original steel section has been lost.

Standard construction practice in the first half of the 20th century was for the voids between the steel frames and the stone and brick masonry to be solidly filled with grout, mortar, or clinker concrete. The grouting of these voids caused stresses created by the expansive formation of rust on the surface of the steel to transfer across the grout into the masonry, resulting in the stone fracturing, becoming displaced and potentially spalling.

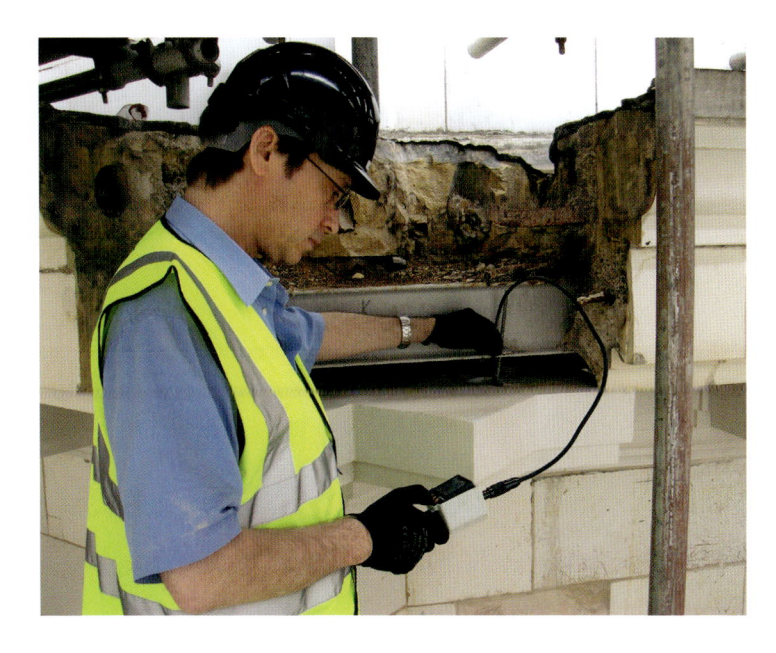

SURVEYS

When surveying a façade built between 1900 and the late 1940's, it is important to look for fractures in locations where the steel beams and columns are known to be or at least anticipated. Usually beams are positioned at or above window head level and columns are positioned at approximate 6m to 9m centres between windows or behind pilasters. Look first in these areas, particularly at high level, under flat roofs, cornices, and balconies. This is where the façade will be most exposed to saturation from rainwater and will be where most of the corrosion will have occurred. Other susceptible locations will be under large areas of glass, and below projecting cornices, particularly where not protected by means of an effective lead or asphalt capping.

LEFT
Water stained Portland stone, Finsbury Circus

ABOVE
Quality control procedures during the reconstruction of faience cornices at the Regents Palace Hotel

137

RIGHT
Portland stone facade, Cadogan Hall

Where the steel is set well behind the stone surface, say at a depth of 150mm or more, it is likely that the condition of the steel frame will be found to be in a good condition. However, where the steel is positioned close to the surface, say at less than 75mm, it is likely that corrosion will be occurring generally across the façade, particularly in the more exposed locations mentioned above.

South and west façades facing the prevailing weather will generally encounter more driven rain than north and east elevations.

Understanding why a steel frame has corroded in one area but not others is fundamental to the development of a successful repair strategy. Repairs involving the exposure and treatment of steel frames are costly due to the necessity to remove and later replace volumes of stone, particularly if the stone is worked or carved. It is therefore important to understand why the steel is corroding and to take steps to remedy defects or improve inadequate details. Without rectifying the defects, the steel will eventually continue to corrode.

Defects typically involve leaking or inadequate rainwater management systems, and inadequate coping or cill details.

Deciding how much of the embedded steel is exposed and treated is not always straightforward. For instance, if a building is occupied, exposing the internal faces of the steel beams and columns might not be practicable. It might also not be physically possible without removing excessive amounts of brick or stonework which could be loadbearing. Exposing the bearing top surface of a steel beam might require the treatment to be undertaken in a sequence of operations similar to underpinning works, or the complete dismantling of the masonry above.

Generally, a level of pragmatism is required. Obviously, severely corroding surfaces should be exposed and treated as a priority. Surfaces facing the internal areas of the building usually corrode to a lesser extent than those facing externally.

STEEL TREATMENTS

Throughout the 20th century several different protective paint systems for treating the surfaces of embedded steel have been used: bitumen, red oxide paint, zinc based paints, and epoxy coatings. The cost of the paint system should not be a primary consideration as this is only a nominal component of the overall repair cost. The general recommendation is to use the best protective system available. For the last 20 years or so, epoxy-based paint systems used to protect steelwork on marine industrial structures have often been considered to be the most appropriate. More recently we have noted that zinc based galvanising systems are being more frequently specified. Adequate preparation of the steel surfaces in accordance with the paint manufacturer's recommendations is essential. Typically this will involve grit blasting though some specifications only require removal of corrosion by means of mechanical wire brushing.

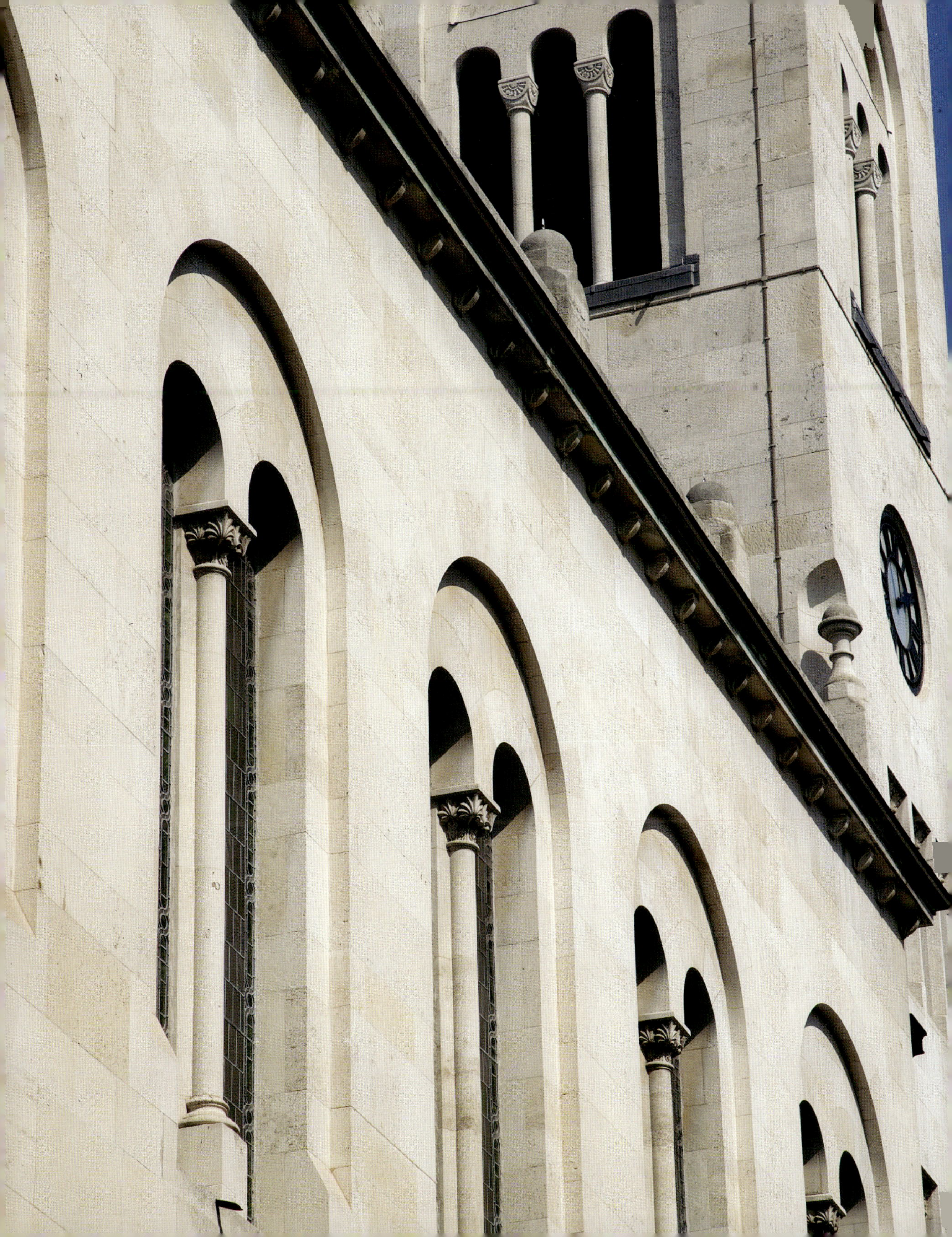

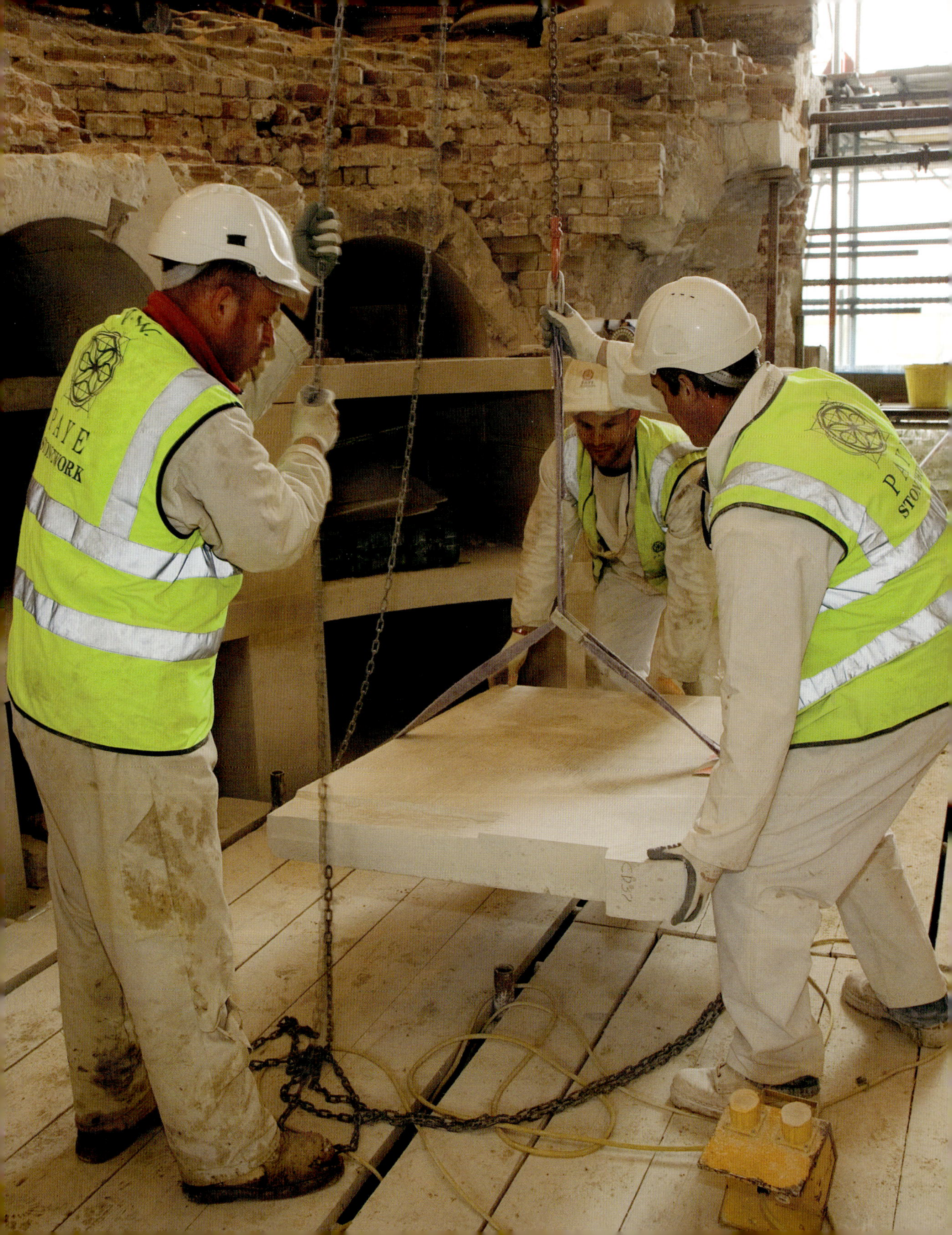
PAYE
STONEWORK

REPLACEMENT MASONRY

Where possible, replacement stones should be the same size as the originals to maintain the original joint patterns. However it is often necessary to form new joints to reduce the volume of replacement stone for practical reasons. It is not usually possible, practical, nor economic to attempt to re-use the original masonry, particularly on columns where the stones need to be carefully broken out because they are toothed and bonded to adjacent courses. Dismantling parapets and salvaging these stones for re-use might sometimes be practical and economic.

If stone is to be salvaged for re-use, a suitable area will need to be made available for temporary storage. Lifting and distribution equipment will need to be provided. The weight of the stone is the primary consideration: its approximate density of 2300kg/m³ (greater if wet) often necessitates the use of a loading gantry.

Replacement stone will need to be adequately tied into the adjacent structure. Often the most practical method of achieving this will be to restrain the replacement stone to the exposed steel frame using straps, cramps, and dowels. Masonry fixings must be stainless steel or other non-ferrous material, whereas the existing steel frame will be a form of mild steel. To avoid potential problems caused by a bi-metallic reaction, an insulating neoprene pad is inserted and stainless steel bolts are sheathed to avoid contact with the mild steel fixing lugs welded on to the steel beams and columns. It is important to undertake all welding before the steel is treated otherwise the paint system will be damaged enabling corrosion to re-occur.

Experience shows us that an embedded steel frame will corrode less where an air gap exists between the steel and the embedding masonry. It is therefore important not to grout the void between the replacement stone and the steel frame.

MAINTENANCE STRATEGIES FOR STEEL FRAMED FAÇADES

Provided rainwater is adequately managed by maintaining the rainwater disposal system to ensure that the masonry never becomes saturated, the repairs have been undertaken to a high standard, and the steel frame is not too close to the masonry surface, the façade should perform well and only need to be inspected every 10 years or so. In marine environments, it would be appropriate to inspect these façades at more frequent intervals due to the increased risks of corrosion.

On façades where the depth of steel frame embedment is less than 150mm it is likely that corrosion will slowly continue in areas not previously exposed and treated

FAR LEFT
Reconstructing the fire damaged crypt in the Darnley Mausoleum

LEFT
Treating corroding steel before reconstructing the faience cornice at the Regents Palace Hotel

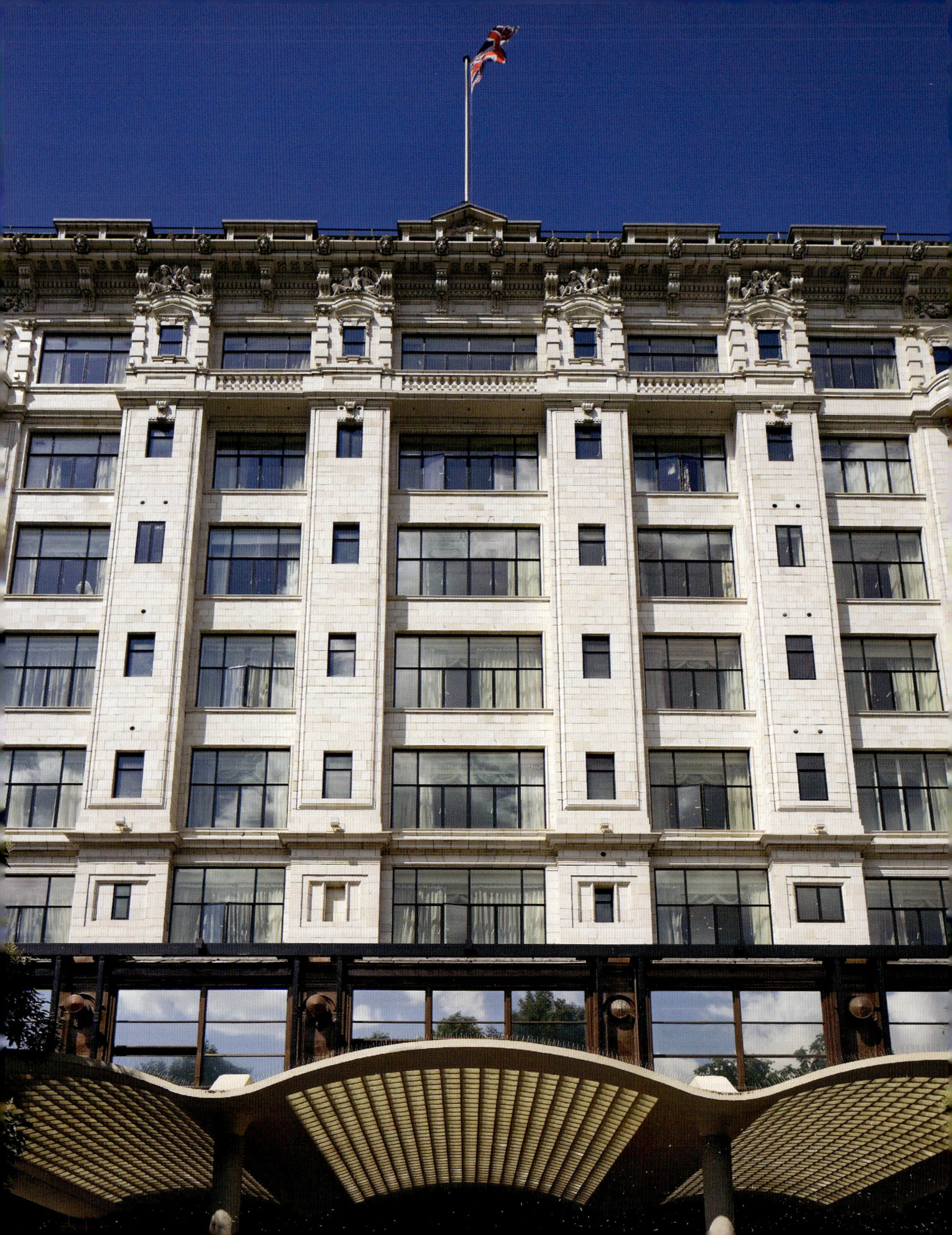

TECHNICAL ISSUES, MANAGEMENT & ORGANISATION

as part of a repair exercise. PAYE have returned to several façades which are maintained on a 10 year cycle. This has given us an interesting understanding of the performance of the façade relative to its quality of design and construction, and the amount of remedial works carried out. For instance, on a Lutyens façade in Fleet Street, PAYE repaired many areas of steel frame in the 1990's and returned 10 years later when the building changed hands to undertake further repairs.

The repair criteria used in the 1990's was that the steel frame would be exposed and treated only where fracturing displacement in excess of 1mm had occurred. Areas where the masonry was only hairline fractured were not repaired. On one façade, no repairs were undertaken due to a requirement to limit expenditure. On this façade, the fractures were merely stitched with stainless steel threaded dowels set in epoxy resin in what could be described as a 'holding exercise'. In the subsequent remedial works, we found that these stitched areas had re-fractured parallel to the original fractures because the stresses had not been relieved by the repairs and so continued to build as the steel continued to corrode. To effect a long term repair it is therefore necessary to relieve the built up stresses and to expose and treat the corroding metalwork.

On façades were the embedment depth is less than 150mm, we recommend that inspections should be undertaken every 2 or 3 years and necessary potentially dangerous defects rectified immediately. Other repairs can be deferred and undertaken as part of the planned maintenance works, perhaps every 10 years or so. Care must be taken to ensure that any displacement and spalling is quickly rectified before it develops into a dangerous condition and threatens the safety of pedestrians below.

Repairs involving the exposure and treatment of the steel frame and replacement of the masonry are often referred to as 'traditional' repairs. The following section deals with an alternative approach: cathodic protection.

CATHODIC PROTECTION

Cathodic protection can take many forms: galvanising, the installation of sacrificial anodes, or protection by means of a permanent impressed current (ICCP).

For any metal to corrode, an electrolyte needs to be present in contact with the metal. In the case of a corroding steel frame, the electrolyte is moisture or rainwater. If charged ions are able to flow through the metal and the electrolyte, an

LEFT
Completed steel frame repairs and cleaning on the faience façades, the Savoy Hotel

ABOVE
Cleaned and repaired Portland Stone façades in Waterloo Place, Westminster

RIGHT
The internal façades of Battersea Power Station awaiting conservation

electrolytic cell will form and corrosion will occur. The impressed current of a cathodic protection system prevents the ions flowing in the manner necessary to cause corrosion, thereby preventing corrosion. An ICCP system involves the use of anodes installed discretely around the steel framed beams and columns. The anodes are connected to a power feed. To ensure that the current is impressed in the correct direction, it is necessary to electrically earth the steel frame. This is where the first problems can arise because there must be full electrical continuity between all members of the steel frame. If the frame is not electrically continuous, corrosion of an isolated steel beam or column might actually be accelerated.

The anodes are usually arranged in zones so that the level of current can be varied from one zone to another. A computer is used to regulate and control the current impressed from the anodes. The anodes are not in contact with the steel frame but installed into the surrounding masonry. It was considered good practice in the 1920's when constructing these façades to grout the voids between the steel frame and the surrounding masonry. However, it is not uncommon to find that these voids have not been grouted, or at least not fully grouted, and that an air gap exists between the steel frame and the masonry. Depending upon the extent of voids, anodes installed into masonry in these circumstances might not adequately impress the current and therefore not protect the steel frame. For the successful installation of an ICCP system, it might therefore be necessary to grout these voids.

The effectiveness of the anodes to impress a current to protect the steel frame should be tested and measured. ICCP costs will be wasted if the anodes fail to do this.

Some low budget ICCP systems are commissioned and left to work without any further control nor adjustment. More sophisticated systems involve regular monitoring, for which there will be a monthly or annual cost. In such systems, the power feed is controlled online through a website. Sensors are installed into each zone to monitor the corrosion potential and the rate of corrosion. The sensors are also monitored on the website. The levels of impressed current can be adjusted through the website as corrosion potential and corrosion rates vary throughout the year.

The cables connecting the anodes and sensors to the computer must be discretely positioned within the masonry joints. This can only be achieved with care.

Further problems arise when a subsequent tenant fails to comprehend the importance of a dedicated online connection and terminates it. We have also known incoming tenants to incapacitate the complete ICCP system by terminating the power feed.

Another important issue which must be fully appreciated is the fact that any metalwork within the anode's 'throw' (which is often up to 1500mm) must be identified and bonded to the steel frame to prevent it corroding at an accelerated rate. Examples of such 'extraneous' metalwork items are rainwater pipes, fixing brackets, window and door hinges, and metal windows.

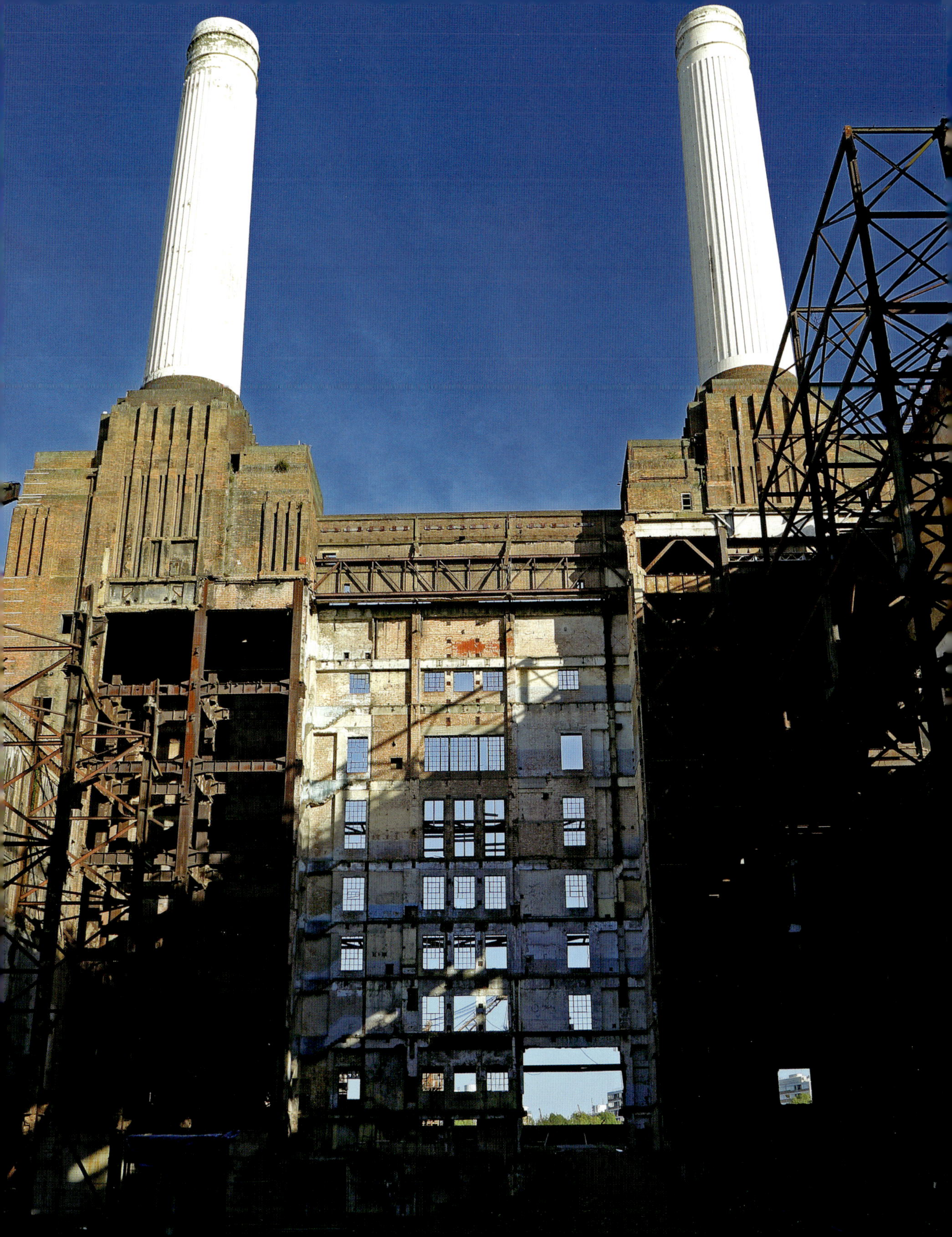

159
Atisreal
shop units to let
all enquiries
020 7338 4000
TO LET
MOSS

Another problem which warrants thorough investigation is the issue of corrosion on the surface of the steel frame. This can insulate the steel from the impressed current if too thick, therefore preventing the ICCP system from working. We have found that more than 1mm of rust will reduce the effectiveness of the ICCP system. In these areas the steel must be exposed and the rust removed for the system to work. Exposing the steel will involve carefully cutting away the masonry, removing the rust and subsequently replacing the stone. Any stresses within the masonry built up by the corrosion of the steelwork will also be removed by this traditional repair.

A cathodic protection system is a means to retard the rate of corrosion of embedded steelwork provided the above risks are properly assessed and appropriate remedial steps taken. If not, the system will fail wasting costs and potentially accelerating the rate of decay. If the steel has corroded to the extent that the masonry has fractured and become displaced, the façade could become dangerous and the priority must be to remove stressed and fractured stone, expose and treat the steel, and replace the masonry. A cathodic protection system cannot make dangerous conditions safe.

Anode manufacturers generally guarantee component life at not more than 25 years.

FAR LEFT
The façades of 159 Fenchurch Street after repair and installation of cathodic protection

ABOVE
Graphite anodes for use at 159 Fenchurch Street

LEFT
Ribbon anodes being installed with reconstructed Portland stone masonry in Cornhill, London

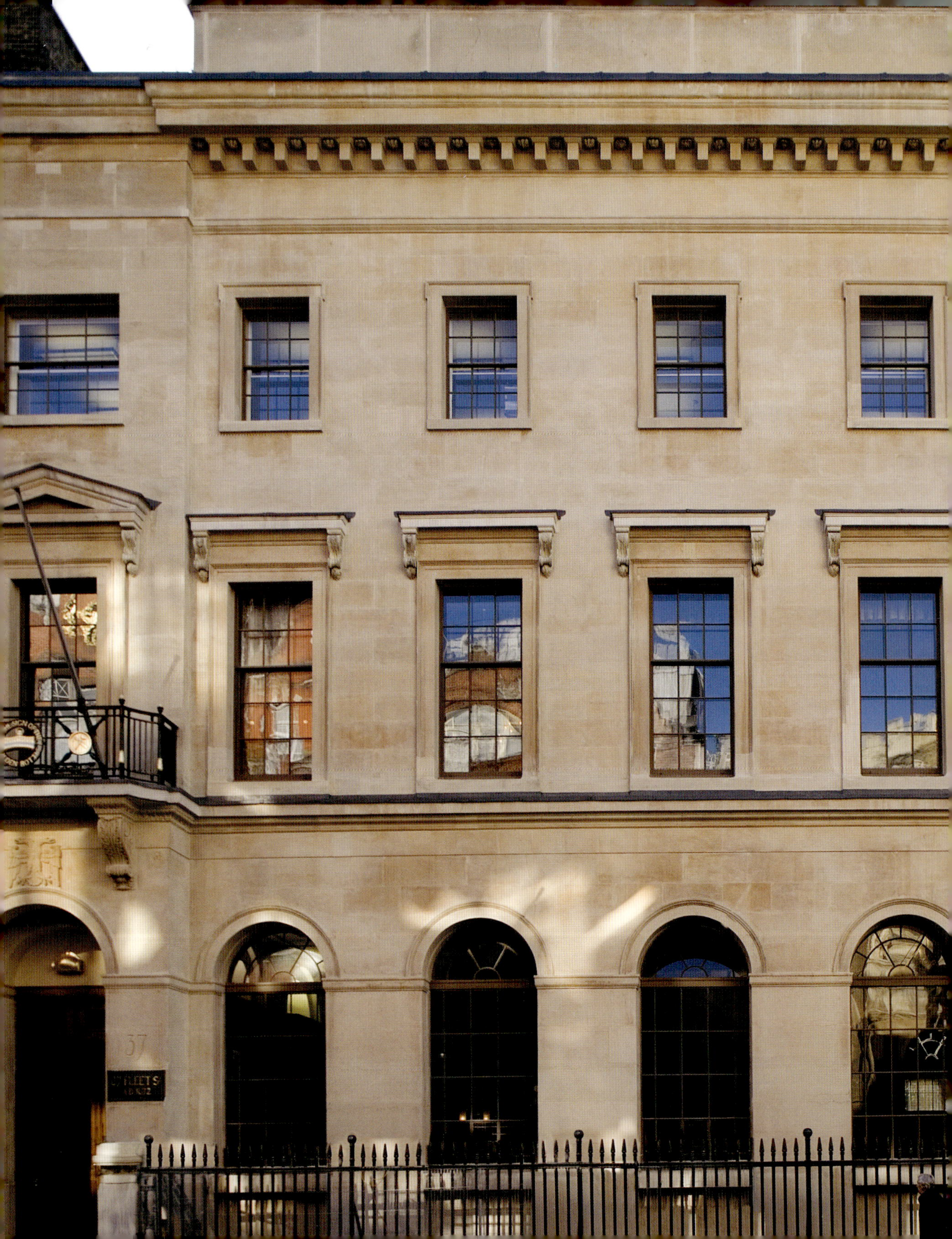
37

TECHNICAL ISSUES, MANAGEMENT & ORGANISATION

Left
Fleet Street Bath stone façade after cleaning and repair with cathodic protection

Above
Installation of anodes in a Bath stone façade, Fleet Street

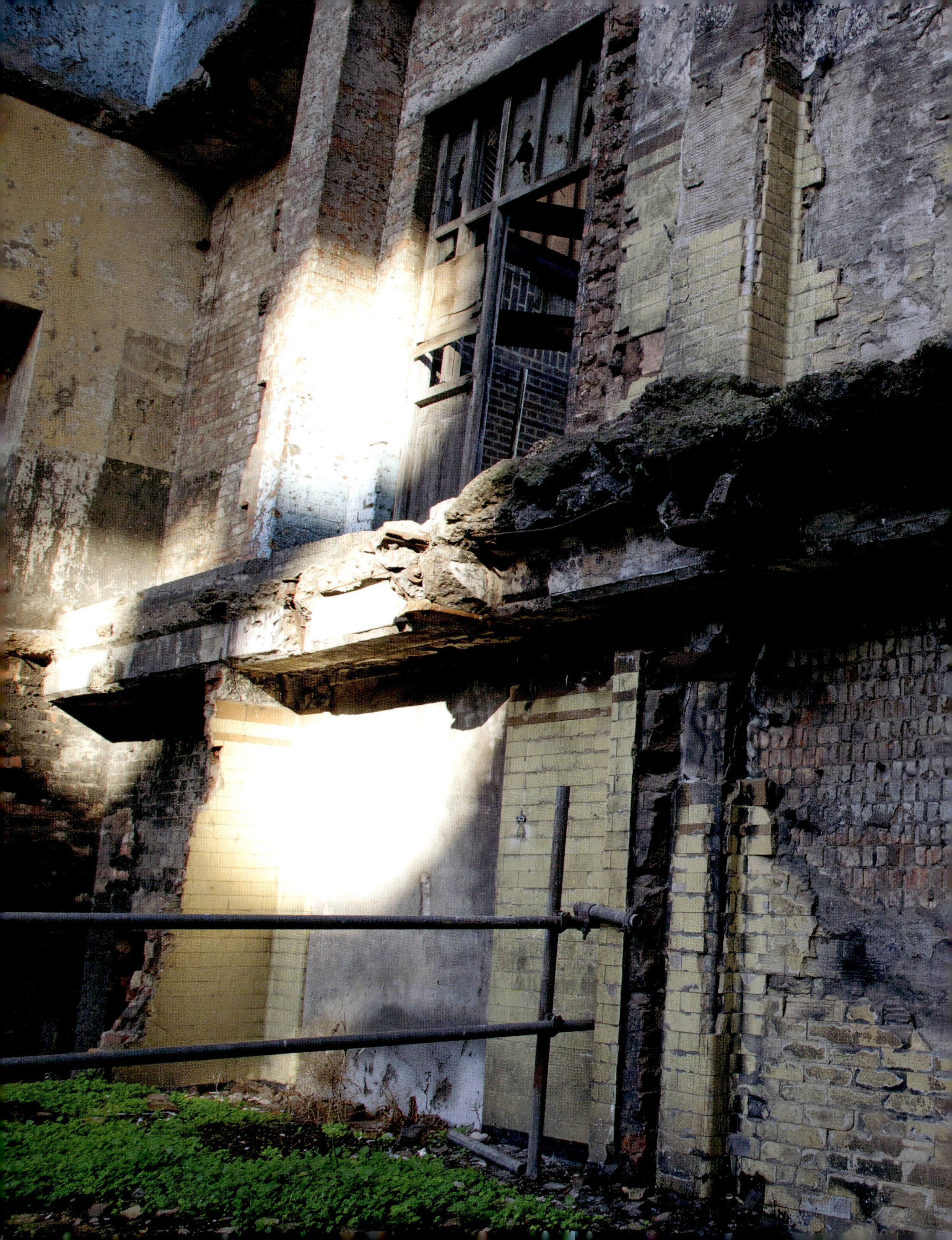

Internal façades of Battersea Power station awaiting conservation

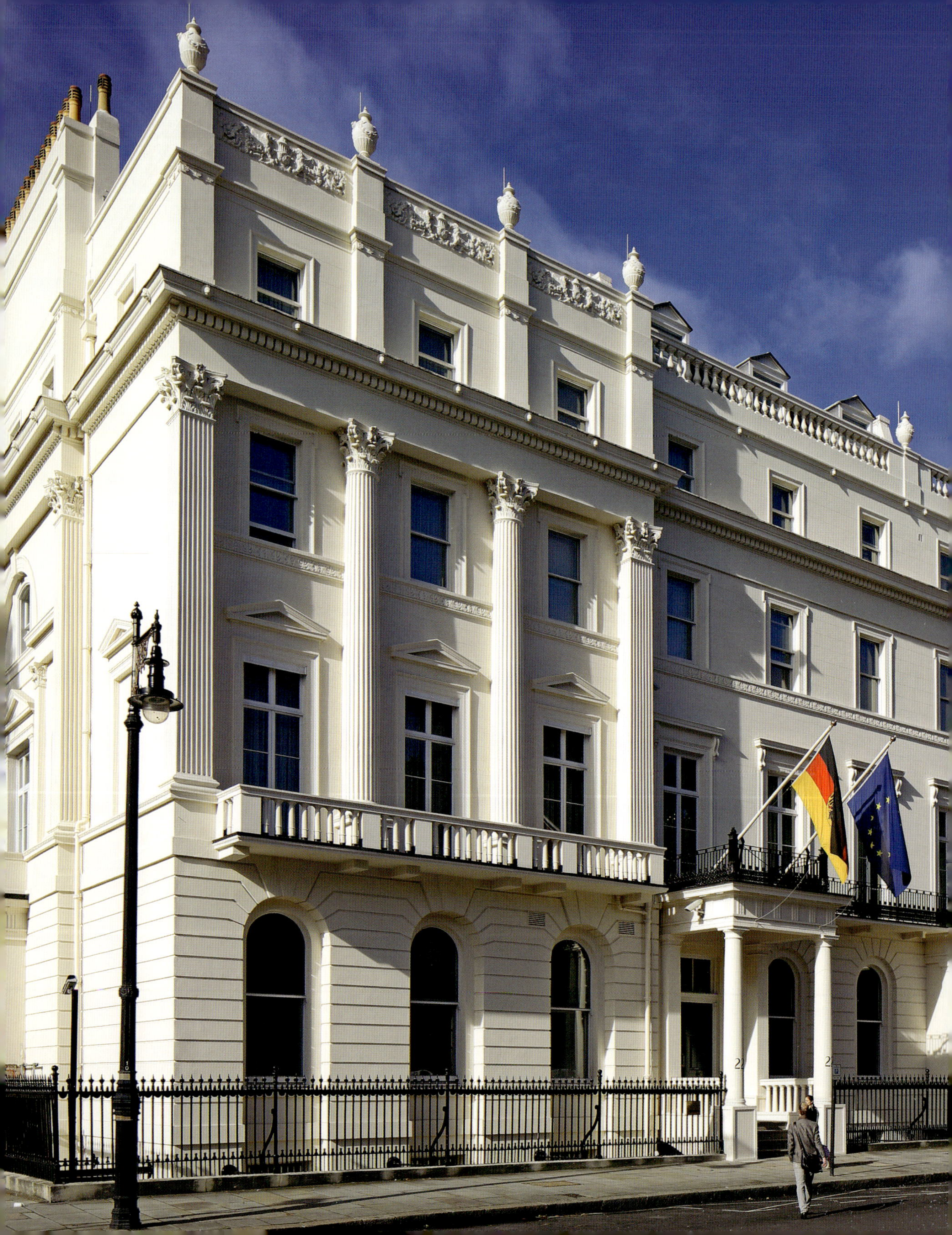

CHAPTER SIX

ALTERNATIVE MATERIALS

RECONSTITUTED STONE

Reconstituted stone is also referred to as recon, cast, or artificial stone. It is a coloured concrete comprised of white cement, crushed stones, other aggregates and sands. It is cast in moulds usually made from good quality plywood, timber or GRP. Rubber moulds are used to form highly decorative undercut shapes. There are two generic types of reconstituted stone: 'dry pack' and 'wet cast'.

DRY PACK uses a very dry mix with barely enough water to hydrate the cement. It minimises the amount of time that the units stay in the moulds (and thus minimises costs as the moulds can be used more efficiently). At PAYE, we are generally cautious about the use of dry packed units. We have found that the concrete mix sometimes does not develop sufficient strength to prevent unavoidable damage handling the units on site regardless of the amount of care taken.

WET CAST uses sufficient water in the mix to fully hydrate the cement but the slump is such that the units need to stay in the moulds for suitably long periods. More moulds are therefore needed than for a comparable project using dry pack mortars making this a more expensive method. However, a superior product is achieved which can also be safely reinforced. Embedded steel reinforcement must be stainless and threaded to ensure that an adequate bond between the steel and concrete is achieved. Wet casting can produce a surface laitance which will craze and this needs to be removed by etching affected surfaces with an appropriate acid after casting.

In the 1970's and 1980's it was common for reconstituted stone units to be cast with a plain concrete core in attempt to minimise costs. The thickness of the artificial stone mortar was a nominal 25mm. Over the subsequent decades it was realised that this was a substandard product as the face debonded from the concrete core. Today, good quality reconstituted stone is monolithic, the same mortar being used throughout.

Reconstituted stone has been used in London from the late 19th century.

The repaired stucco render façades of the German Ambassador's Residence in Belgrave Square

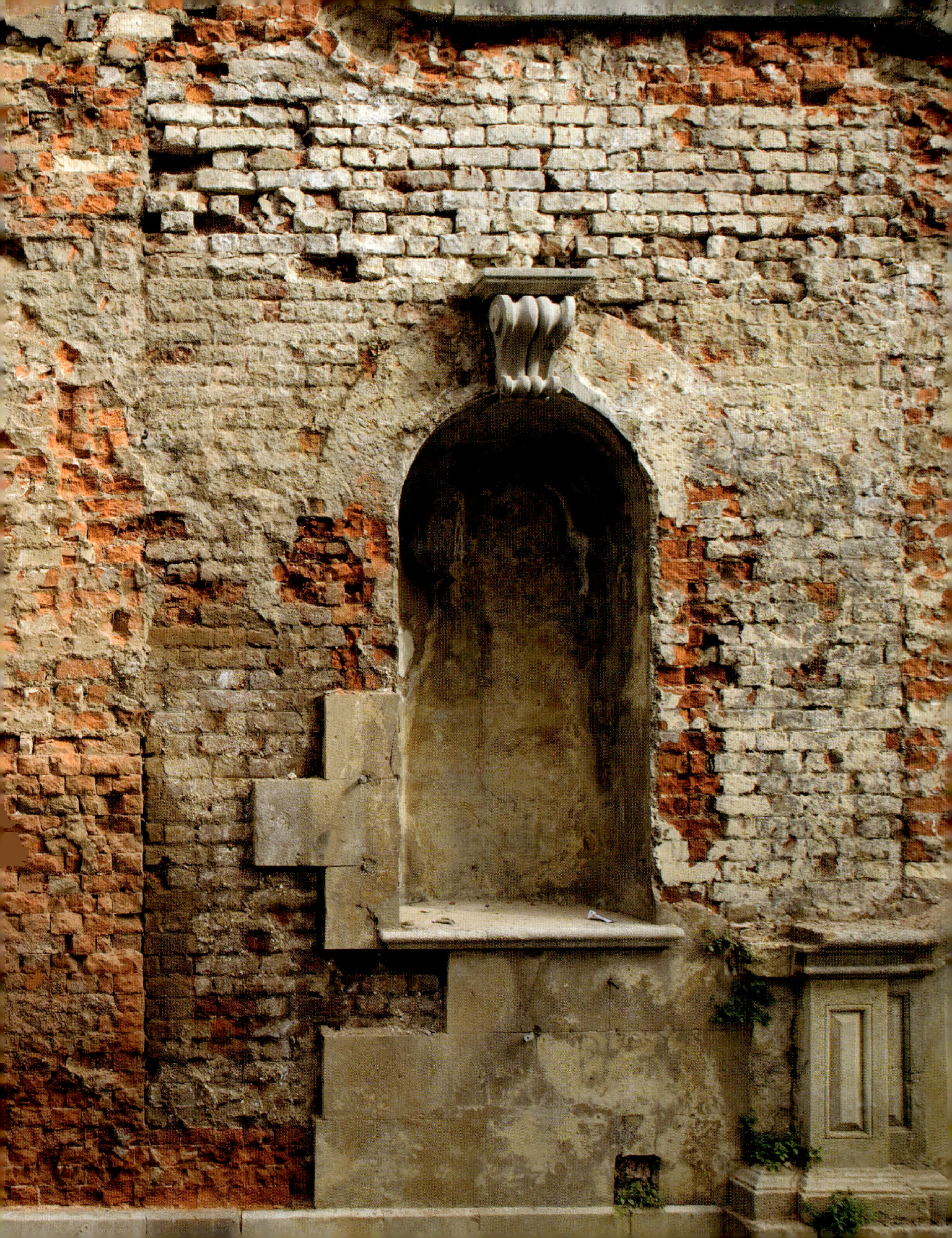

ALTERNATIVE MATERIALS

On early examples it was commonly used in conjunction with natural stones where visible at close range, for instance at ground floor level around entrances and shop windows.

Early reconstituted units were heavily reinforced with plain mild steel, often surprisingly complex in its 'cage' design. Predictably, this reinforcement corrodes causing the reconstituted stone to spall particularly where the stone is more exposed to rainwater saturation. Today, reinforcement is stainless steel and threaded to ensure a mechanical bond is achieved with the cast stone concrete mix.

In London it is common for Portland stone to be replicated and this is achieved using white cements, Portland stone dust, and aggregates which tend to be either a crushed white limestone, or a dark flinty aggregate. Where a dark aggregate was used and surface erosion has weathered away the binding white cement, the appearance of the weathered reconstituted stone becomes more that of the dark aggregate. No amount of cleaning will be able to return the reconstituted stone to its original colour, and client's expectations need to be carefully managed. This problem is less of an issue where white limestone aggregate has been used.

BRICKWORK

BRIEF HISTORY

The earliest bricks were baked in the sun and used in Egypt approximately 15,000 years ago. The earliest kiln-fired bricks were used in Mesopotamia from about 6,000 years ago. Bricks were extensively used throughout the Roman Empire in a variety of shapes but in England they were typically long and thin, usually less than 38mm high. After the 5th century collapse of the western Roman Empire, brick use in western Europe all but ceased. The Saxons and, later, the Normans are known to have salvaged and re-used the durable Roman bricks, an example being St Albans Abbey in the 11th century. Brick manufacture remained in the Byzantine Empire, Islam, and parts of Italy from where it re-

FAR LEFT
Exposed brickwork on the south terrace of Cliveden House uncovered by the removal of defective 19th century Roman cement stucco render

ABOVE
Stucco repaired and cleaned at Kenwood House

LEFT
19th century Roman cement stucco rusticated ashlar detail at Cliveden House

ALTERNATIVE MATERIALS

Demand in England for brick flourished in the 15th century following its acceptance as a sufficiently prestigious façade material by the church and the aristocracy.

Frogs appeared in bricks in the late 17th century to enable quicker drying and firing. They made the brick lighter and helped to 'key' the mortar.

In the 18th century, the quality of bricks improved due to legal statutes being imposed which defined standards. This in turn lead to the development of better kilns.

Famous examples of the historic use of brick include the following examples, all of which have been repaired by PAYE: Herstmonceux Castle (1440), Hampton Court Palace (1515), St James's Palace (1531), and Kew Palace (1631).

MANUFACTURE

Brick clay used to be prepared and pressed by hand into moulds. After the invention of the steam engine in England in 1769, the preparation of the clay and the pressing of the clay into moulds became mechanised, and by the mid-19th century over 100 different brick making machines had been patented.

After pressing, the bricks would be removed from the mould and left to dry slowly for a period of weeks, then fired in a kiln. For the first few days of firing, the temperature was kept low, then slowly raised to between 800 C and 1100 C before slowly cooling. The firing process took between 15 and 30 days depending on the size of kiln, the fuel, and the weather. Modern kilns substantially improve these periods.

Above
Chenil House brick façade after hydraulically jacking into new position 3500mm further forward from original

Right
Red brick and Ancaster stone at St Pancras Station

emerged, later returning to England from the Low Countries in the late 13th century.

Unsurprisingly, brick manufacture and use flourished in areas such as Flanders, the Netherlands and northern Germany where stone was either not available or difficult to extract and transport. The first bricks used in England (after the Romans and Norman's re-use of Roman bricks) were imported from Ypres in the 13th century, including over 200,000 bricks used at the Tower of London in 1278.

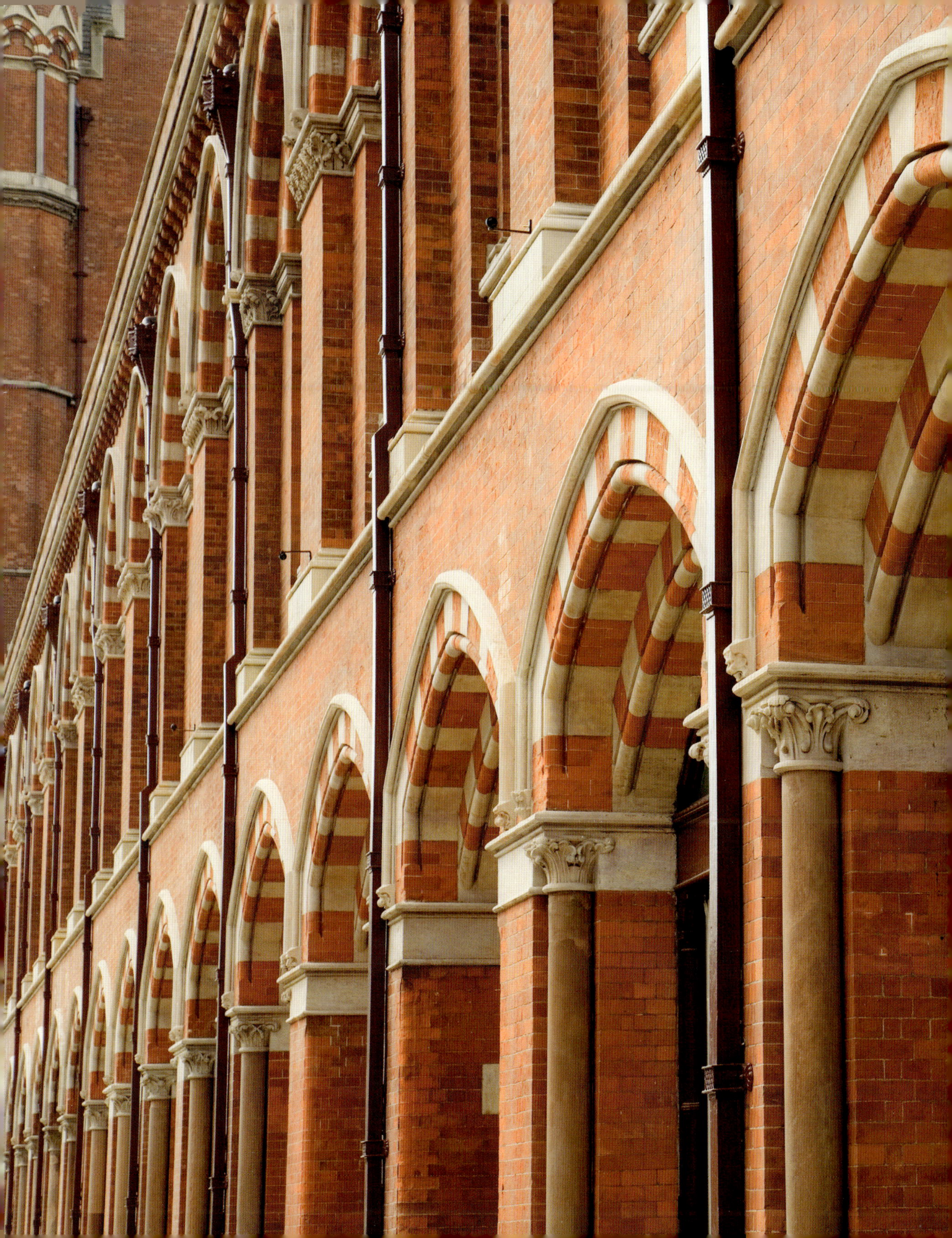

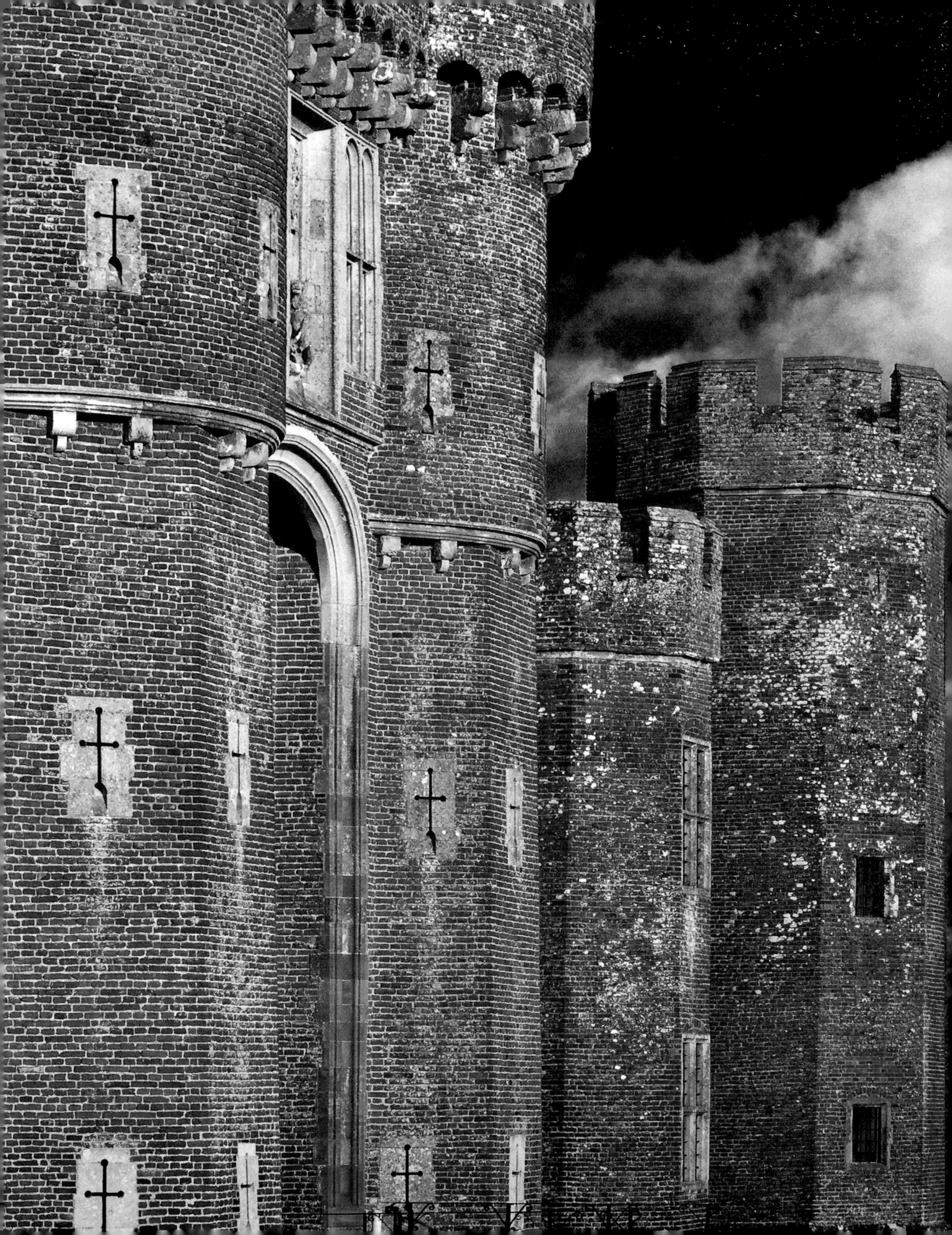

ALTERNATIVE MATERIALS

Modern bricks can either be hand pressed, machine pressed, or extruded and wire cut.

The firing temperature affects the colour of the brick. In early kilns the variable temperatures achieved in the kilns resulted in variable colours. The traditional Kentish brick quilt pattern arose because the shape of the Kent kilns, and the placing of the bricks within them, caused the header faces to be fired at higher temperatures resulting in the headers having a darker colour than the stretchers.

Colour is also affected by minerals such as iron oxides and calcium within the clay.

CLASSIFICATIONS

Facing bricks are made to achieve a consistent aesthetic for facing façades and must be of good appearance, strength, and durability. They should be free from surface damage and manufactured to a specified dimensional tolerance.

Engineering bricks are manufactured to achieve a specified crushing strength (typically either 50 or 70 N/mm^2) and minimum absorption (typically either 4.5% or 7%).

Common bricks are those suitable for all uses other than for facing or engineering bricks.

BONDING

Early important buildings would have been constructed by immigrant Flemish craftsmen who fully understood the accepted rules and practices. However in the majority of other buildings, the bonding was haphazard, and the need to adequately bond bricks together was not properly understood. Gradually, the knowledge and appreciation of the correct bonding procedures became established.

Brickwork constructed in London from the mid-17th century is regarded by many as some of the finest anywhere. Gauged brickwork, using soft accurately rubbed bricks with fine joints of only 1mm or less, was introduced from Holland.

Common bonds are Flemish and English, both with garden wall variants. Rat-trap bond was often used on rural estates for farmworker's cottages and reduced costs by minimising the number of bricks required per unit area.

LEFT
15th century red brick at Herstmonceux Castle, Sussex

ABOVE
Sloane Telephone Exchange

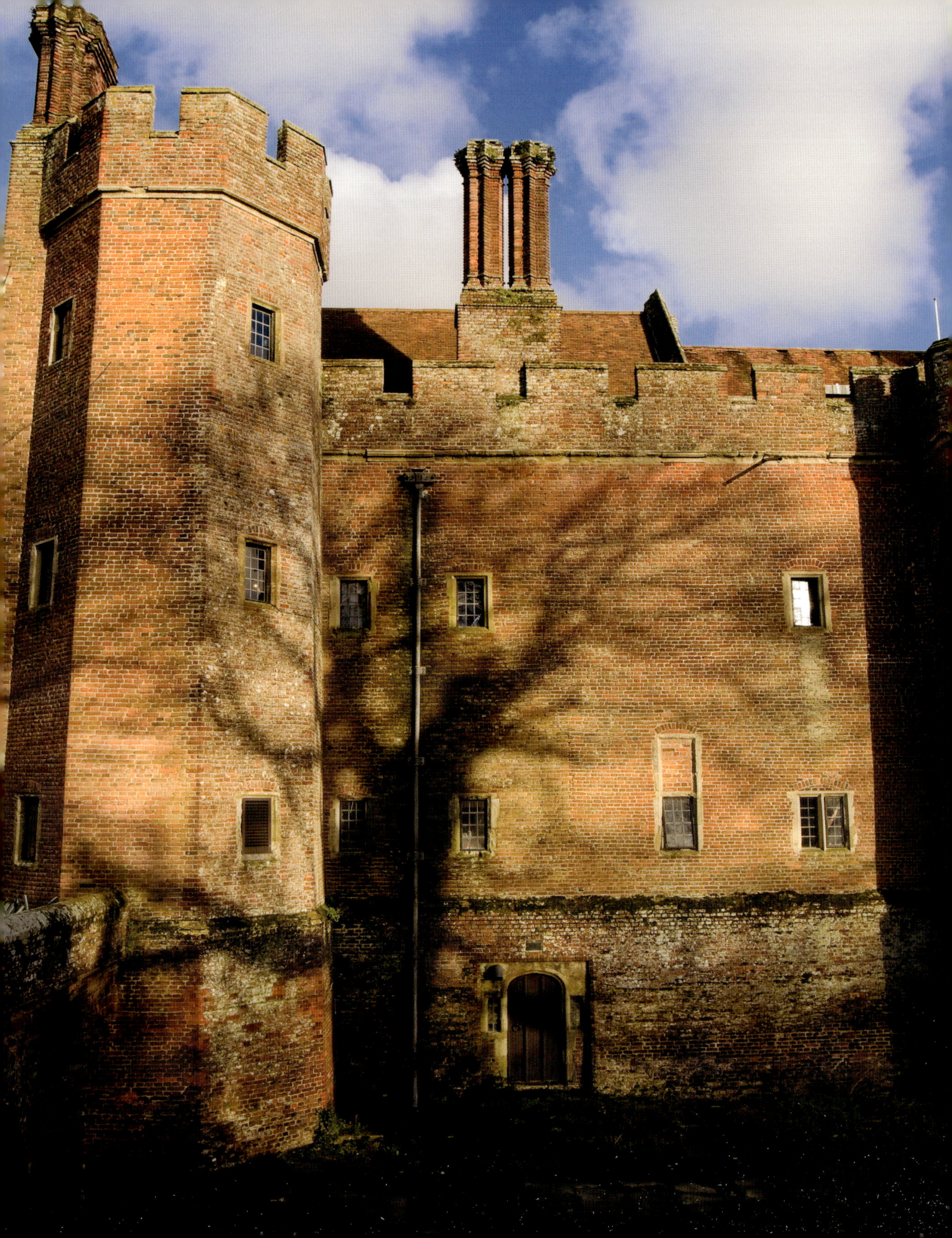

ALTERNATIVE MATERIALS

DEFECTS

Well-fired bricks with low sulphate content are relatively durable and will perform in exposed locations for long periods of time. Under-fired bricks will be less durable. Over-fired bricks will be smaller, darker and more easily broken.

Most well-fired historic bricks have good frost resistance but some bricks were manufactured specifically for non-external use at low cost and these will obviously fall if used inappropriately.

Frost damage occurs where rainwater is absorbed then later freezes and expands. Ice crystals grow bigger than the brick pores in which they form and mechanically break down the surface of the bricks. Frost resistance is variable. For example, red bricks often fare worse than yellow stocks.

Salts can be naturally present in the clay and these are generally chlorates or sulphates of calcium, magnesium, or sodium. These salts will remain in the clay during the firing process and have the potential to dissolve in rainwater and migrate towards the surface where evaporation occurs. The salts re-crystallise where this happens and this has the potential to damage the brick surface but more often just appears as an unsightly rash. Most salts are eventually removed by the action of the wind and rain, others have the potential to harden and become permanent disfigurements on the brick surfaces. If salts form they are best removed by repeated brushing using a dry brush taking care not to damage the surface. A weak solution of acetic acid can also be used but it should be noted that any wet treatment has the potential for the solution to be absorbed by the brick.

It is not uncommon to find that the external facing brick skin has been constructed independently of the backing brickwork and has no mechanical bond. This is particularly common where gauged bricks have been used.

FAR LEFT
Red brick and sandstone at Herstmonceux Castle

ABOVE
Red brick at Eastbury Manor

LEFT
Sulphate pollution on red brick at Battersea Power Station

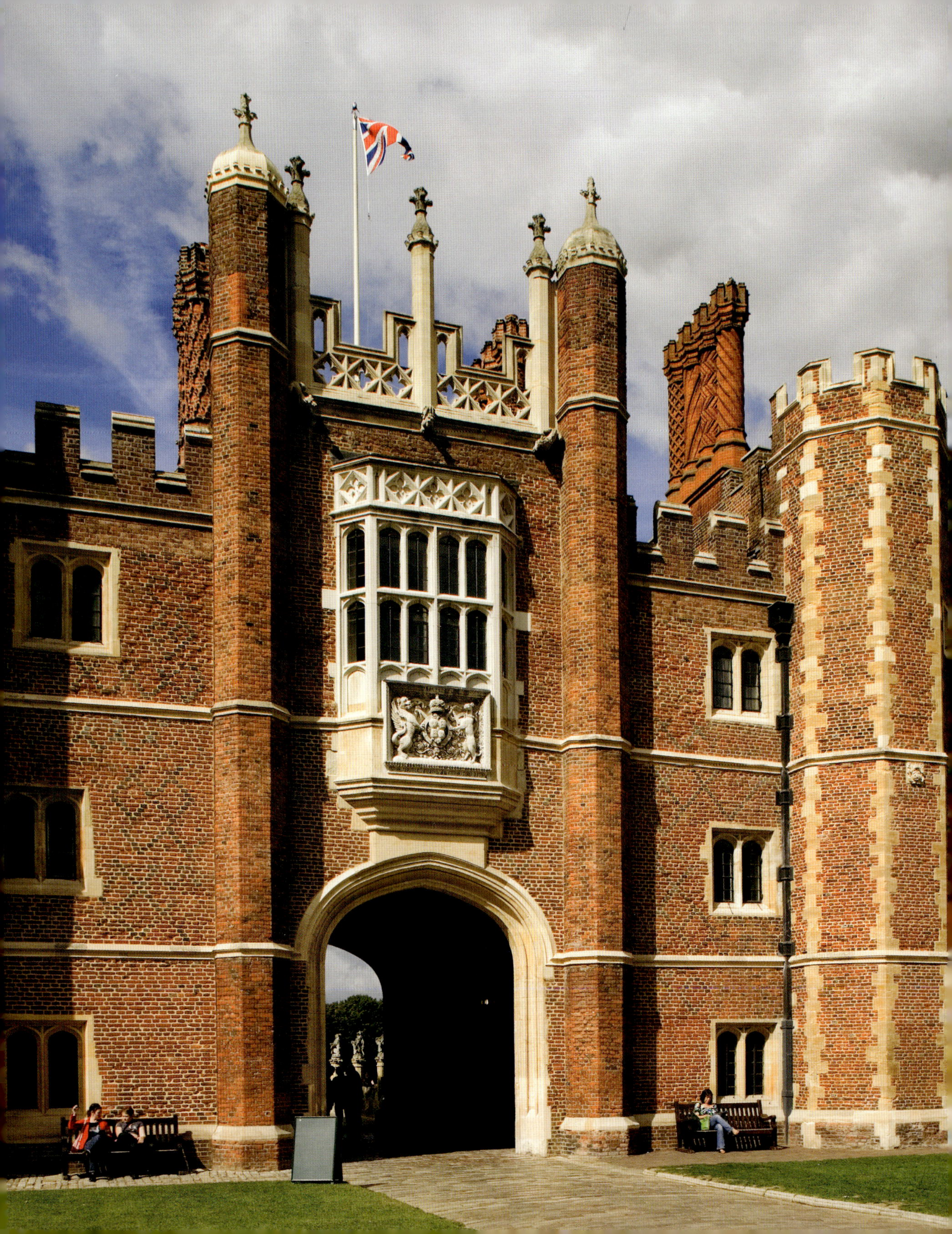

ALTERNATIVE MATERIALS

16th century red brickwork at Hampton Court Palace

REPAIR

Attention to detail is necessary for the successful repair of historic brickwork.

This will typically involve:
- The identification and rectification of the cause of the problem. Often this involves maintaining or improving a poor rainwater management detail. Occasionally the problem is merely exposure to weathering and poor durability.

- The replacement brick should match the original in terms of colour, texture, size, dimensional tolerance, porosity, and shape. It is usually necessary to manufacture purpose made bricks to achieve this. Historic bricks were made to non-standard imperial sizes and surprisingly tight tolerances. Modern pre-manufactured metric bricks will not suit an historic façade because the dimensions inevitably will not match and current standard tolerances are too generous.

- The existing bond must be matched.

- The existing joint width must be matched.

- The existing mortar must be matched in terms of colour, texture, style, and finish. It is usual for the aggregate to be exposed as a final treatment within a day or two of the mortar being applied. It is never appropriate for the pointing of historic brickwork to be finished with a metal tool and a weatherstruck profile.

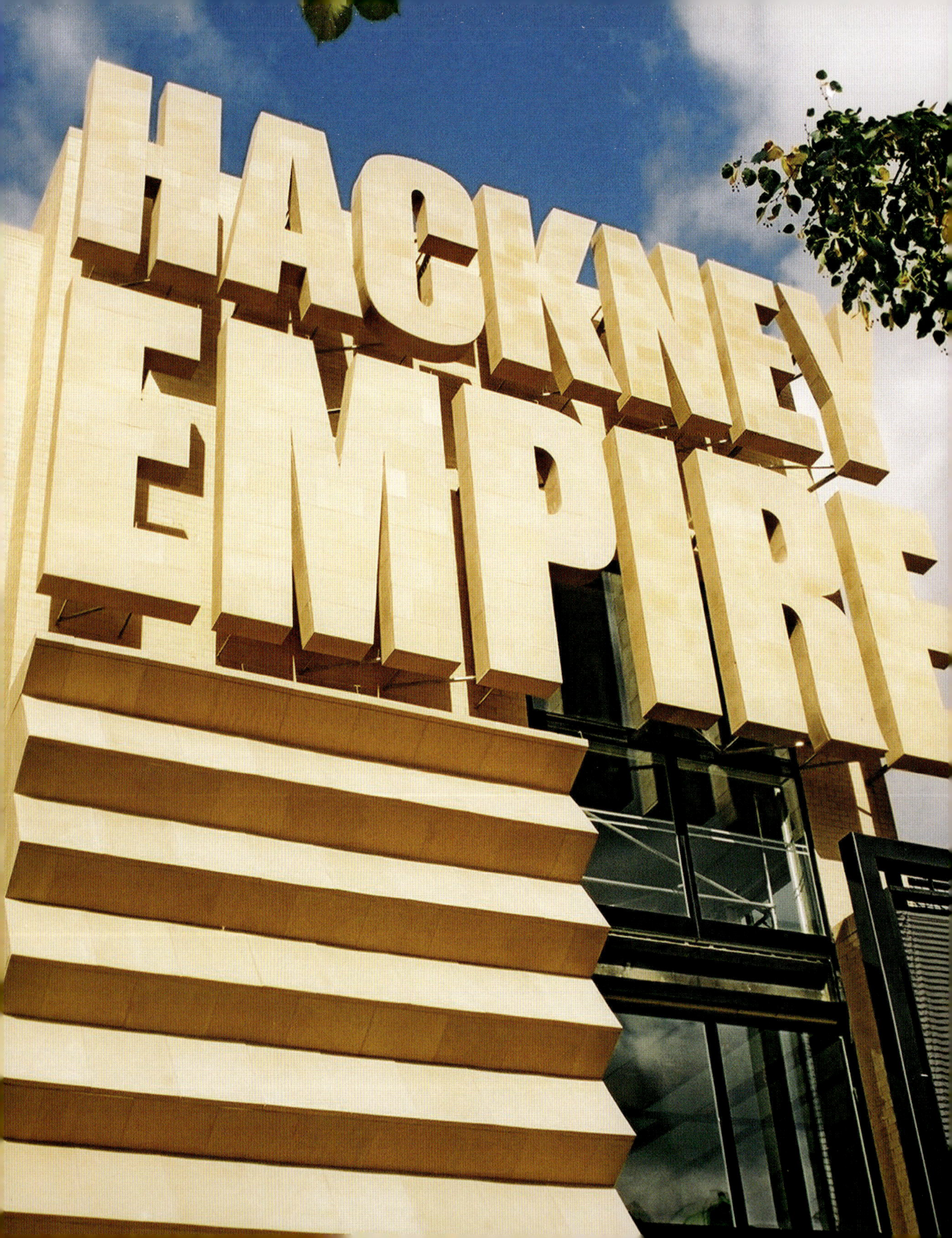
HACKNEY
EMPIRE

ALTERNATIVE MATERIALS

TERRA COTTA AND FAIENCE

Terra Cotta is an early form of artificial stone made by casting clay in moulds and kiln firing the cast blocks to produce relatively durable hollow units. Its general use dates from the 1850's though an earlier form was produced by the Coade family in London from the mid 1780's.

Terra cotta blocks were made by companies such as Carraraware, Burgantoffs, Doultons, Hathernware, and Shaws of Darwen.

In a UK construction context, 'faience' is the term used to describe terra cotta which has been glazed.

The production of architectural terra cotta and faience reached a peak in the 1880's and continued well into the 20th century with the available ranges of colours and moulds particularly suiting the art nouveux and art deco architectural styles. By the 1930's it was being used generally as a faience tile, typically 38mm or 25mm thick. These were often bedded against the backing brickwork using adhesive cement mortars without any mechanical fixings.

Traditionally, the faience glaze would be applied to fired units and re-fired. Many of today's glazes have been developed for application onto the cast clay before the first firing, thus eliminating the need for a second firing.

Terra cotta hollow masonry blocks are cast with a wall thickness of up to 35mm. They became increasingly popular because of their ability to withstand the harmful effects of heavily polluted city atmospheres. Clear salt glazes were commonly used in the north of England and white glazes became popular in the south, particularly in London in the early 20th century.

The hollow blocks were filled by the masons during construction using whatever materials were to hand. Usually this was a loose mortar of broken bricks but occasionally clinker concrete was used. In the projecting elements of cornices, heavy filling materials were avoided and anything from wood shavings, old ropes, or newspaper was used instead.

The function of the filling material has been much debated. Current consensus is that the filling did not provide much additional compressive strength but merely acted to provide extra mass where it was needed and to secure any restraint fixings embedded into the hollow blocks. Technical

ABOVE
Clinker concrete and twisted rope filling in fractured terra cotta blocks at the Regents Palace Hotel

LEFT
The completed new terra cotta façades at the Hackney Empire in 2004

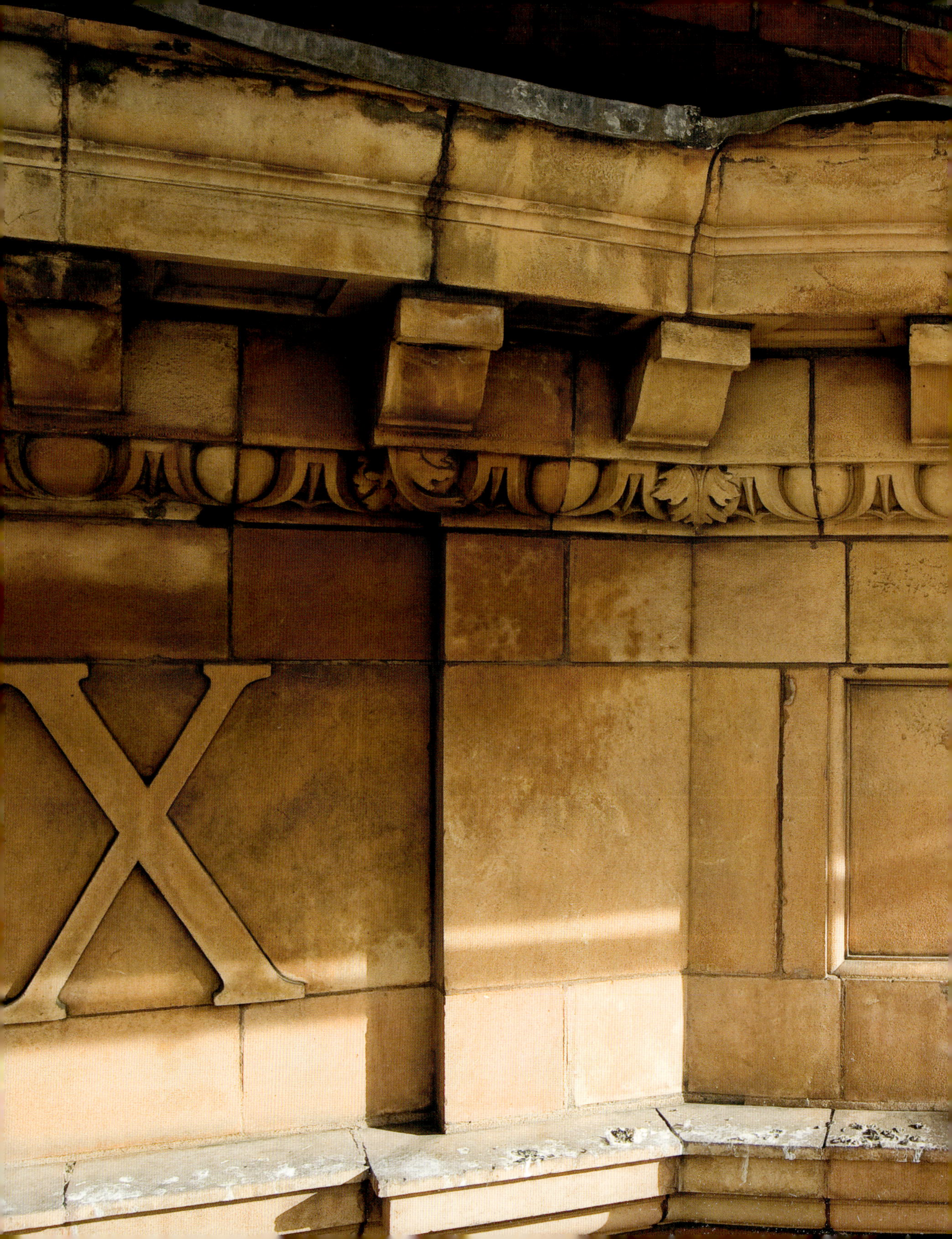

ALTERNATIVE MATERIALS

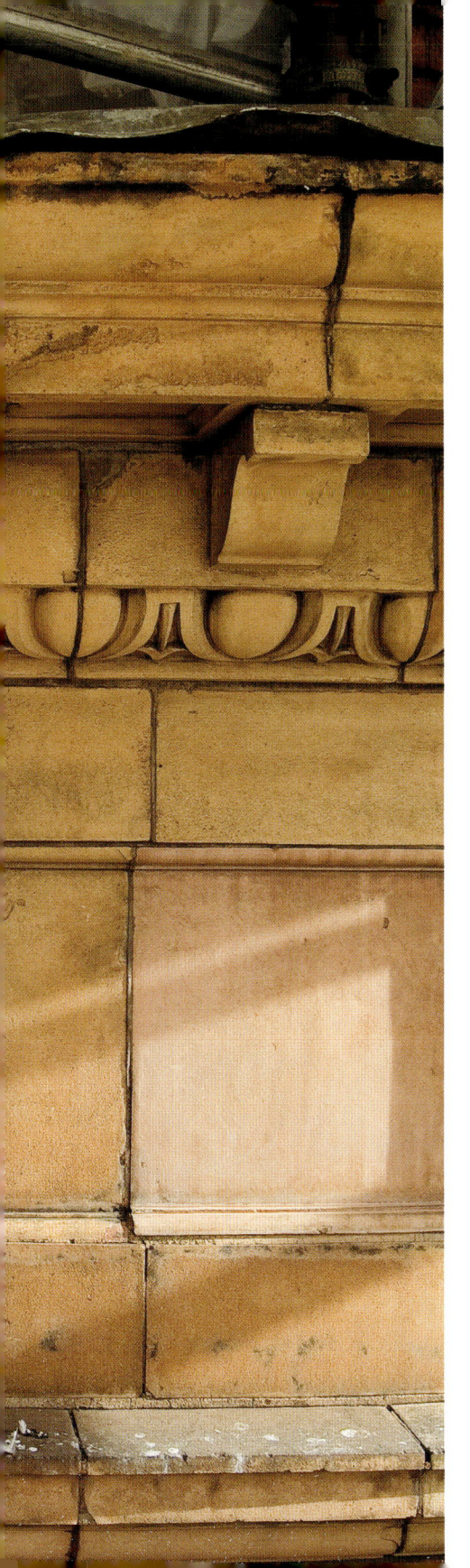

Terra cotta cornice detail at the old Middlesex Hospital during conversion to residential accommodation

advice from competent manufacturers is that the filling should not be compacted into the hollow blocks to avoid the risk of damage from exothermic reactions within the mortars.

In order to replicate the large projecting cornices of natural stone façades, terra cotta and faience cornice units needed to be suspended by means of cramps, ties and dowels from steel cantilevered beams. Much damage has been caused by these corroding, and the soffits of cornices are often the first places to show signs of failure where corrosion is occurring. These defects are also potentially dangerous: any masonry spalling from the soffit of a cornice is likely to fall to the pavement below. Corroding cramps causing fractures will often be present in other exposed locations on the façade such as parapets and copings.

Many cornices were constructed with inadequate weathering details involving an unprotected gutter on the sky surface drained by means of inadequate small diameter lead pipes which easily block with debris and cause the cornice to become saturated. It is common to find many failed attempts to prevent rainwater finding its way into these cornices. At PAYE, we have found the best long term solution is to adapt the falls above the cornice to allow rainwater to run away from the building line and to drip off the cornice, and to install properly detailed leadwork protective cappings with suitable drips and upstand flashings.

Terra cotta fired at a lower temperature than usual will appear more as a salmon pink colour and this is usually found to be less durable than the more typical buff colours. Terra cotta develops a skin on

ALTERNATIVE MATERIALS

Above
Terra cotta façades of the Whitechapel Gallery featuring new artwork by Turner prize winning artist Rachel Whiteread Commission: Tree of Life (2012)

Right
Installing new faience at the Regents Palace Hotel

Far Right
Terra cotta and brick details at the old Middlesex Hospital during conversion works

its surface during firing and it is this that provides the resistance to weathering and pollution damage. If this skin is weakened as a result of under-firing or the use of inappropriately aggressive cleaning methods, the terra cotta unit will become susceptible to accelerated decay.

PAYE have analysed many of the mortars historically used for the bedding and pointing of this material and found that 1:3 cement and sand was generally used, equating to 1:4 using modern OPC or a 1:1:6 cement, lime and sand mortar.

A characteristic of many historic faience glazes is their tendency to craze. This usually occurs over a long period of time and is caused by the creation of stresses as the clay substrate expands and the glaze contracts. The resultant fractures have the potential to damage the clay substrate if the stressing is excessive, and this can act as a point of failure for water to locally saturate and degrade the block. Generally, however, surface crazing is not an issue that affects long term performance.

Another issue that can cause a detrimental effect is expansion of the clinker concrete filling material which will occur when rainwater is able to penetrate the block and it becomes saturated. Clinker concrete uses semi-burnt coal as an aggregate and this expands as rainwater is absorbed, fracturing the terra cotta unit.

Terra cotta and faience blocks and tiles are still made in England and elsewhere in Europe. The manufacturing procedures are such that the lead-in times are measured in months.

HOSPITAL

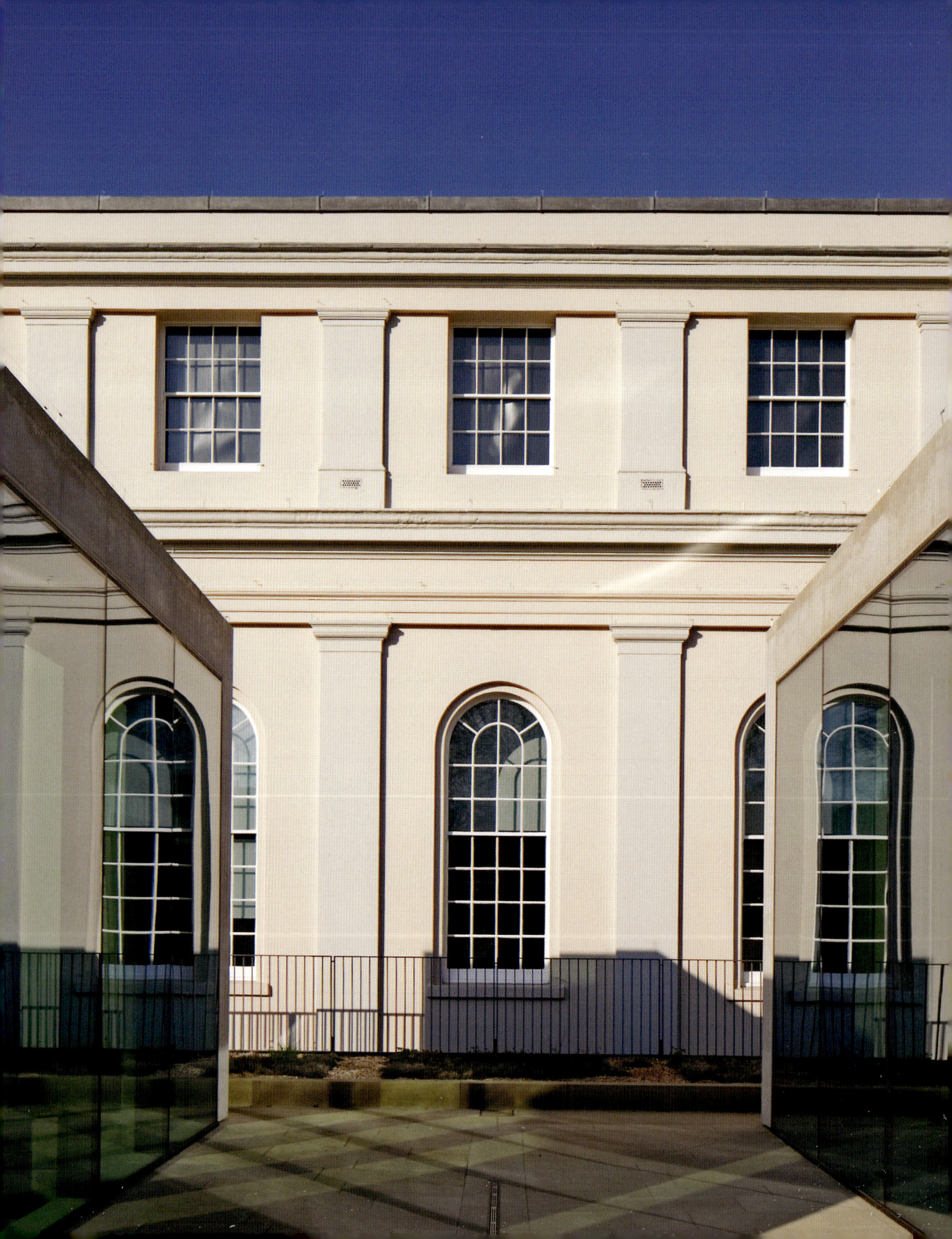

ALTERNATIVE MATERIALS

Modern gas-fired kilns produce units of a consistent colour whereas historic terra cotta and faience were produced in coal-fired kilns of inconsistent temperatures providing a consequential variety of colours. It is therefore usually deemed necessary when replacing units upon an historic façade to produce new terra cotta and faience in a variety of suitable colours to match the original range.

STUCCO RENDER

BRIEF HISTORY

Painted stucco render on brickwork was used as a low cost alternative to stone, typically in speculative residential developments in areas such as Holland Park, Lancaster Gate and Notting Hill where short term profit was the principle requirement rather than long term low-maintenance performance.

Stucco façades built from the late 18th century well into the 19th century used an aerated, adhesive and durable mortar known as Parker's Roman cement. This was patented in 1796 and was made from naturally hydraulic septaria extracted from the Thames estuary at Northfleet in Kent, fired, then crushed and mixed with sand. The stucco was traditionally sealed with six coats of boiled linseed oil and decorated with oil based paints.

Decorative moulded courses such as cornices and window surrounds were formed by hand using profiles cut to the required shape. Individual details were often cast in moulds and adhered to the stucco substrate using the adhesive Roman cement. In later façades, the pre-cast elements were often secured using cast-in metal straps. Were present, these are now usually found to be corroding.

Large projecting cornices were constructed over cantilevered York stone slabs built into the brick façade. The height of the cornice was built up from the York stone corbel using brick and tile make-up, before being rendered in three or more coats to the required moulded profile.

LEFT AND ABOVE
Stucco repairs and redecoration at the Royal Naval College, Greenwich

171

Detail from the Whitechapel Gallery showing new terra cotta 'negative' window casts and tumbling gilded leaves by Rachel Whiteread Commission: Tree of Life (2012)

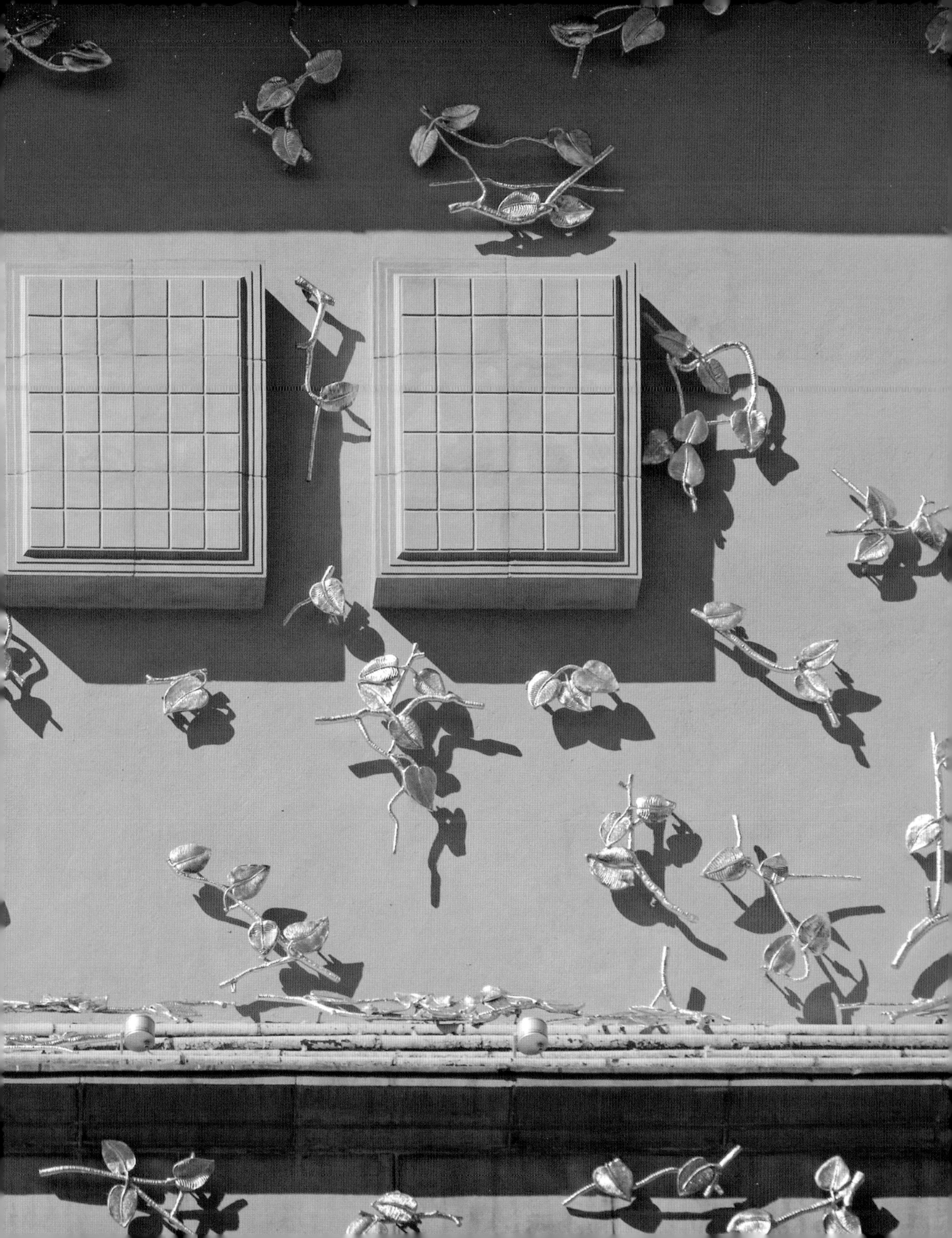

ABOVE
Cast Roman cement finial details, Hadlow Tower

RIGHT
Cast Roman cement details and stucco render on stainless steel framing, Hadlow Tower

DEFECTS

The projecting elements of a stucco façade tend to suffer most from the detrimental effects of repeated saturation from rainwater, particularly the sky surfaces of cornices and copings. In extreme cases the brick and tile make-up work fails in addition to the decorative render. It is unusual for the York stone corbel to fail, though this has been known. The installation of protective lead cappings over these features will dramatically reduce future maintenance costs, as would the installation and maintenance of properly detailed rainwater management services.

Previous repairs involving the use of cement mortars tended to fail within the short term, and it is not unusual to now find façades repaired using these mortars to need all previous repairs to be replaced.

Window openings are usually spanned using a half brick thick arch facing two or more backing timber lintels. The timber lintels are often found to have rotted, resulting in the brick arch becoming overstressed and failing.

It is not uncommon to find dry rot under the stucco render in the backing lintels, roof and floor timbers.

Many of these façades have poor foundations causing vertical fractures to appear, typically running between window openings over the full height of the building.

REPAIRS

Many stucco façades have been subjected to very poor repairs over the decades, and to bring a façade back to its original appearance usually involves extensive work. For example, old patched render repairs are often found to be finished at the surface level of the paint, which itself can be many millimeters thick. If the paint is to be removed, these repairs will stand proud of the original surface and either need to be 'feathered in' or replaced.

Removing the paint will reveal all previous repairs and underlying defects. Not only will these defects need attention but the paint removal itself will often require weeks of painstaking, disruptive and messy work. It was common for the paint to be removed using blow torches but, as this damages the stucco surfaces and creates obvious fire risks, this practice

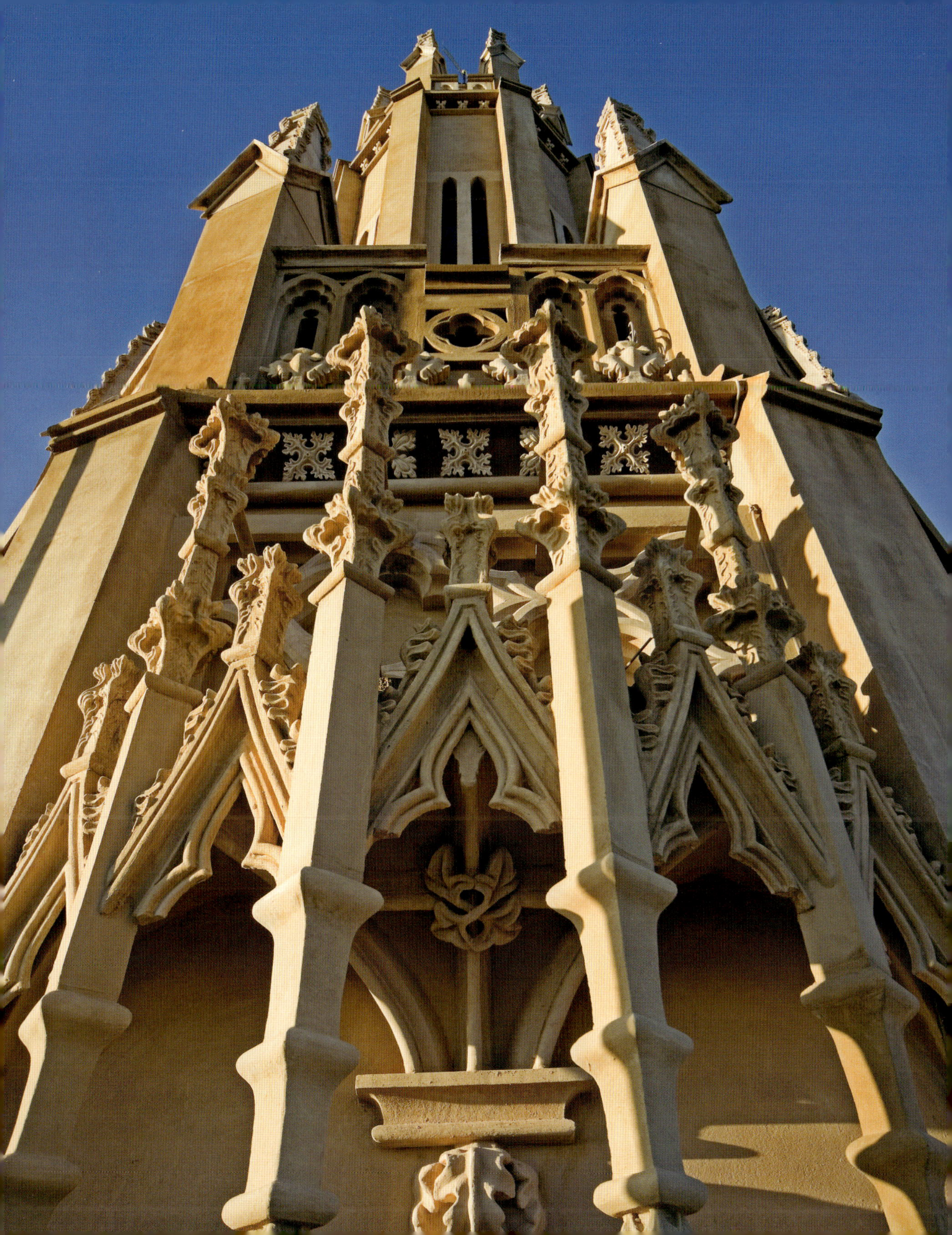

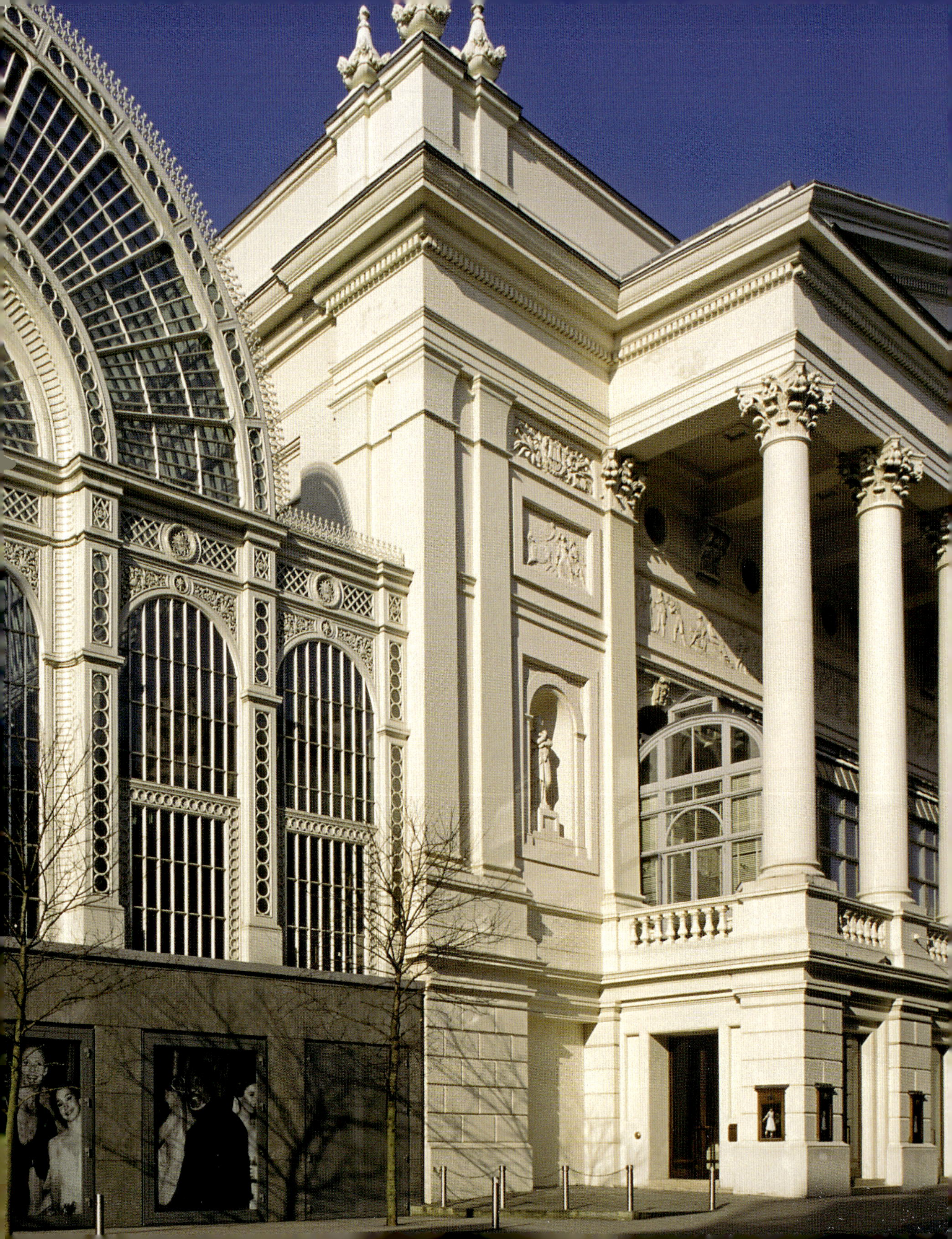

ALTERNATIVE MATERIALS

should now never be used. Any form of grit blasting to remove the paint must also be avoided because of the damage that will be caused. The most appropriate method by which paint can safely be removed is to use poultices washed off by means of pressurized hot water. This is a laborious job particularly where the stucco render is deeply embellished. Some paint stripping poultices have the potential to mobilize salts either pre-existing or left as a residue after poulticing (or a combination of both). Such salts can have a detrimental effect on any subsequent paint system so thorough testing is always a necessity, as is taking manufacturer's advice at an early stage.

Mortar mixes need careful consideration. Current preferences are to avoid modern cement mortars, however, as Parker's Roman cement is no longer available, care needs to be taken in specifying a suitable alternative. The French company, Vicat, produces a natural hydraulic cement which they describe as Roman cement, but this is very different from the original Parker's Roman cement. In their Building Conservation series of books, John and Nicola Ashurst suggest various mortars for repairing stucco render including 1:1:6 and 1:2:9 cement, semi-hydrated lime and sand mixes and 1:3 hydraulic lime and sand mixes. At PAYE we have found that all mixes benefit from some of the 'sand' component being crushed stone dust. We have achieved very good results using all these mortars, however, caution is necessary using lime mortars in the autumn because of the long curing periods and the need to maintain the mortar at minimum temperatures for a period of months. This is difficult if the scaffolding has been removed after completion of the works.

If using lime mortars in the autumn or winter it will be necessary to fully encapsulate the work and protect it from rain. The temperature of the render should be maintained at a minimum 15 °C, ideally using propane heaters which create both moisture and carbon dioxide to assist in the curing process. The curing period will be extended by 25% for every 5 °C drop in temperature. Most lime mortars will need about 3 months to achieve a degree of frost resistance at an ambient temperature of 18 to 20 °C and relative humidity of 55 to 80%.

To achieve a sound stucco repair, substrate fractures will need to be repaired and stabilized. This might involve replacement of fractured bricks or the installation of stainless steel stitching rods provided all movement of the façade has ceased. If movement is ongoing, advice will be required from a suitably qualified structural engineer.

It is good practice to undercut edges where new stucco abuts existing, and to use stainless steel EML screwed into the substrate spanning 150mm either side of any old fractures.

The 19th century stucco façades of the Royal Opera House with Coade stone frieze salvaged from an earlier building

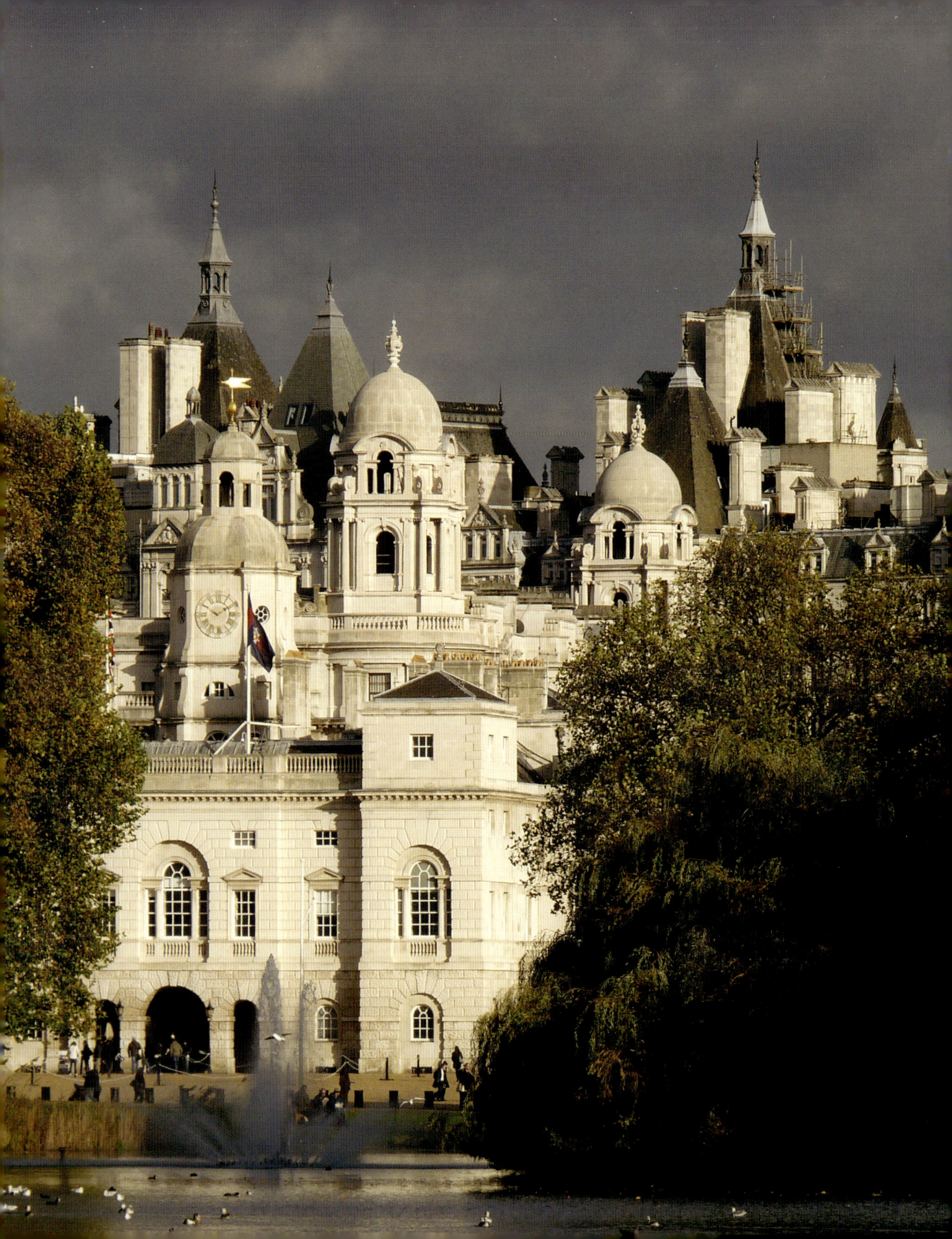

CHAPTER SEVEN

A STONEMASON'S HISTORY OF LONDON

When asked to identify a piece of stone, the first step is usually to identify the age of the building. This is because specific stones were used at specific stages throughout London's 2000 year history, and knowing where a building sits on this timeline will help to identify the likely materials and technologies. This timeline is what has been described below.

Most of the information is not available online but has been assimilated over decades undertaking conservation work on thousands of façades. Pieces of the jigsaw have been found in a castle built more than 900 years ago, the careful dismantling and reconstruction of an important historic façade, the story behind an old photograph, or a relevant sentence in a book.

This story concerns economics, politics, power, and the development of London's infrastructure. It can not be completed because there are too many unanswered questions and gaps in our knowledge, but new facts do come to light and these slowly help us to better understand and explain the past.

This is a stonemason's history of London. Other cities will, of course, have other stories.

THE FOUNDING OF LONDON

In 47 AD the Romans chose to establish London in its strategic river position on adjacent areas of raised ground now occupied by St Pauls Cathedral and the Bank of England and separated by a stream known as the Walbrook. In this location the river Thames was narrow enough to bridge but deep enough to allow ships to dock in the tidal waters. The Romans were superb engineers and introduced technologies not seen before in Britain: the use of levers, winches and arch formers; the use of kiln fired clay tiles and bricks; and most importantly, the use of mortars.

Mortars enabled stones to be bedded and adhered together, not only providing resistance to wind and rain but enabling masonry structures to be built efficiently minimising the width of the wall, and hence the necessary volume of stone. The maximum height to width ratio of a wall built without mortar will be in the region of 2.5:1, whereas that of a wall built with

Horseguards Parade, government offices, and hotels in Portland stone, Westminster

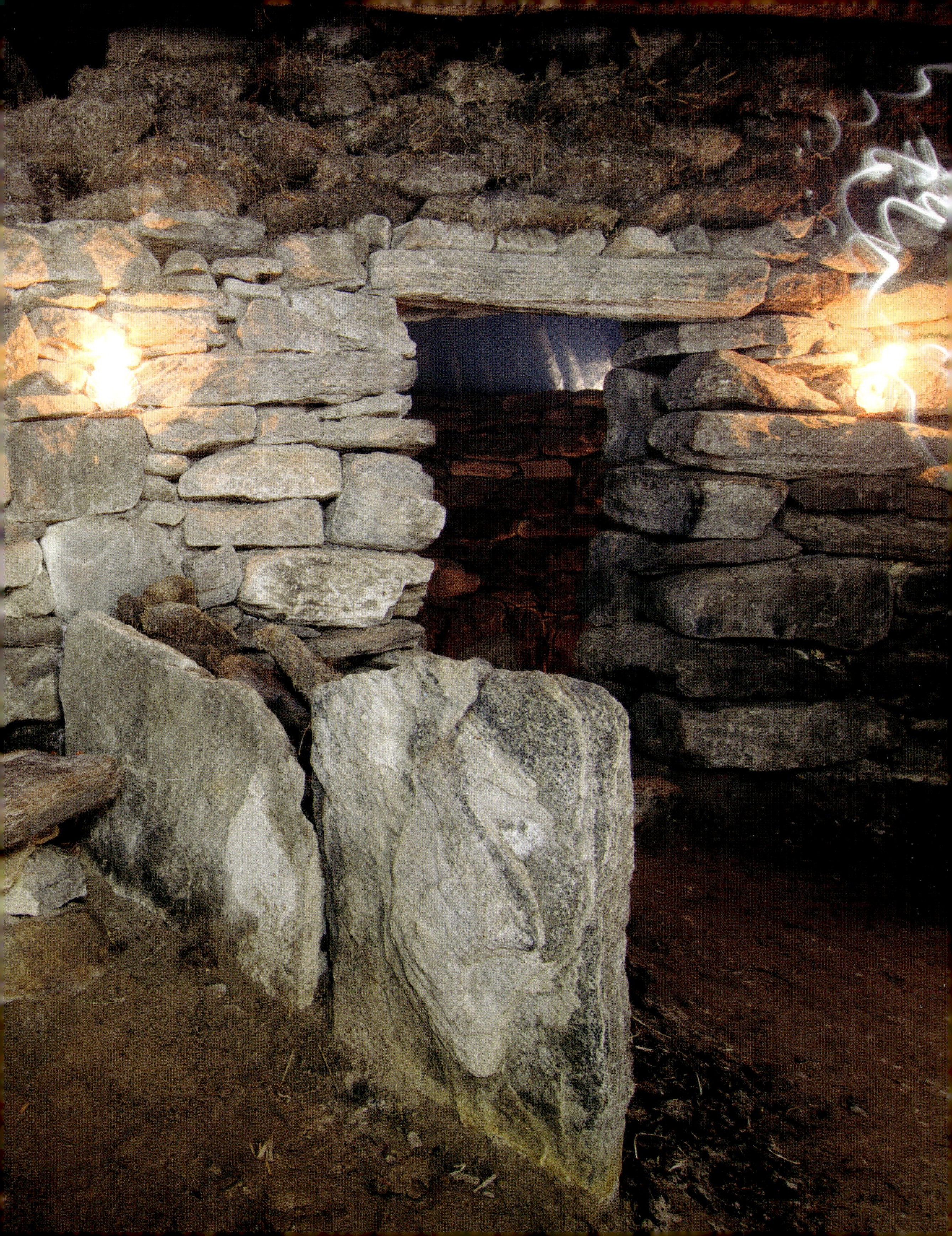

mortar will be closer to 10:1. In other words, a 75% saving in the volume of stone could be achieved using mortar. In London this was particularly important because stone was not available locally and needed to be brought in by barge. This was obviously expensive in terms of labour and resources. Until the industrial revolution and the construction of the railway system, transporting heavy stone by water was the only practicable means.

Pre-Roman wall construction in Britain was either timber framing with wattle and daub, or in areas where timber was not available, stone bedded in earth. The difference in technical abilities between Roman and early British masonry façade construction can perhaps best be demonstrated by comparing a photograph of Roman masonry built with mortar in Pompeii before the eruption of Mount Vesuvius in 79 AD to that of the 'dry' masonry walls of a surviving but restored 3000 year old dwelling on the Isle of Lewis and similarly constructed Scottish black houses which were being built until the late 18th century. It seems that mortar was not readily available on the Isle of Lewis until the latter half of the 18th century when this form of construction ceased (being replaced by brick with lime mortar). The slender and efficient Roman walls feature windows and arches whereas the 'dry' British walls are windowless and squat with low roofs.

The nearest available supply of stone which the Romans quarried for constructing their city of London was Kentish Ragstone, quarried near Maidstone in Kent and transported by barge down the river Medway and up the Thames. The remains of a Roman barge were found in 1962 during construction work on the north bank at Blackfriars. It had sunk with 25 tons of Kentish Ragstone on board. Remains of the barge have been conserved and can be seen at the Hastings Shipwreck Museum. It has been calculated that 1,750 loads of Kentish Ragstone would have been required for the city defensive wall, construction of which commenced in 190 AD. Many more loads would have been required for dwellings, temples, military and administrative buildings. The remains of a Roman amphitheatre exist under the Guildhall.

LEFT
Dry stone walling in a domestic building in the Hebrides dating from nearly 3000 years ago

TOP
Prehistoric steps and stone walling set in earth in the Hebrides

ABOVE
Black houses built in Scotland using stones bedded in earth

181

A STONEMASON'S HISTORY OF LONDON

London was a significant trading centre. Its forum was the largest Roman market north of the alps, but compared to other British Roman cities its buildings were simple. To date no decorative columns nor capitols have been found, nor have any large temples. Whether this simplicity can be attributed to the nature of the Kentish Ragstone, which is difficult to work into complex shapes, or expediency is not known.

AFTER THE ROMANS

The western half of the Roman Empire collapsed in 410 AD. The Romans abandoned their city of London which, evidence suggests, had been in decline for the previous 150 years. Interestingly their technologies ceased too, perhaps suggesting that the knowledge was not shared with the indigenous British. Construction methods reverted to those of the pre-Roman times: no bricks were produced in kilns, and no masonry was constructed with mortar. This low level of technology continued for centuries.

A new city slowly emerged outside old Roman London in the area now known as Aldwych. This was mostly constructed in timber and very little remains now.

The earliest masonry dating from after the Roman occupation in London has been found at the church of All Hallows Barking in Great Tower Street. An arch exists which is difficult to date but is thought to have been constructed in the 8th or 9th centuries. Roman bricks and quarried stone have been re-used to construct this opening. The Roman ruins would have been used as a readily available source of stone. It seems that the only masonry built in this period was for churches and abbeys, and nothing

RIGHT
Roman masonry in Pompeii in AD79

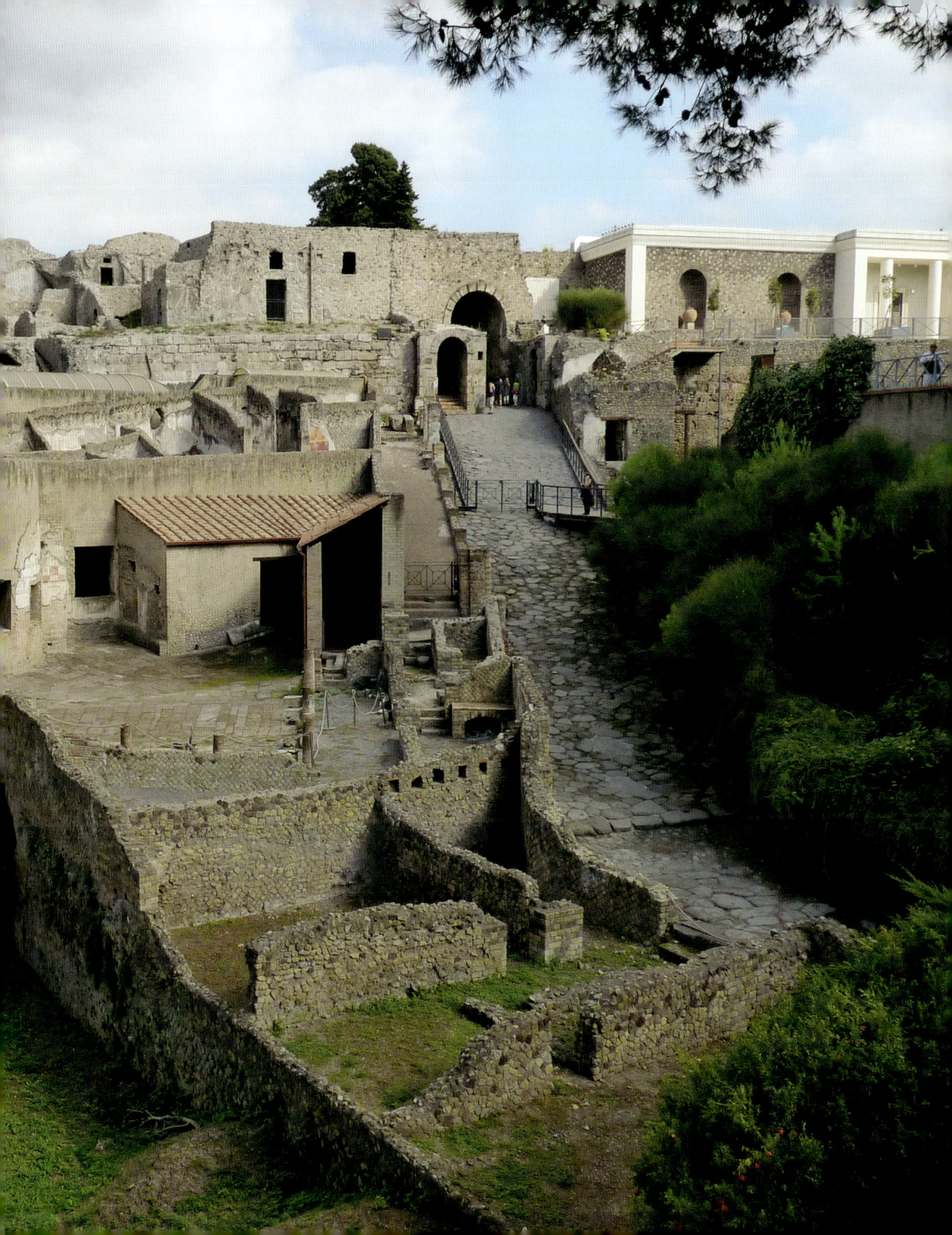

Above
Roman bricks forming column profile in Pompeii in AD79

Right
Roman brick column with render to replicate masonry in Pompeii in AD79

but their foundations now survive. It is interesting to consider why this should be so: could it be that the skills (and knowledge of lime mortars) were only available for ecclesiastic buildings, and were the skills imported from elsewhere (probably from the technologically superior Christian Frankish empire across the channel)?

Another impediment to a greater use of stone in this period might have been that the Saxon kingdoms were small and did not possess the military might of the Roman Empire to ensure that stone could be safely quarried and transported relatively long distances without substantial economic risk.

In the century before the Norman invasion of 1066, the construction of abbeys and churches increased but there are many references to failures. For instance, Ramsey Abbey was constructed in 968 but condemned and demolished less than thirty years later. It is not surprising that little remains from this period.

Causes of failure probably include insufficient foundations, bad workmanship (then referred to as 'scamping'), and the poor understanding, preparation, and use of lime mortars.

Any churches from this period still standing in London were destroyed by a hurricane in 1091.

THE IMPACT OF THE NORMAN INVASION

Norman skills in military warfare are widely recognised. Less so are their extraordinary skills as stonemasons.

Decades before the Norman invasion of Britain in 1066, the Normans had become expert stonemasons as demonstrated by

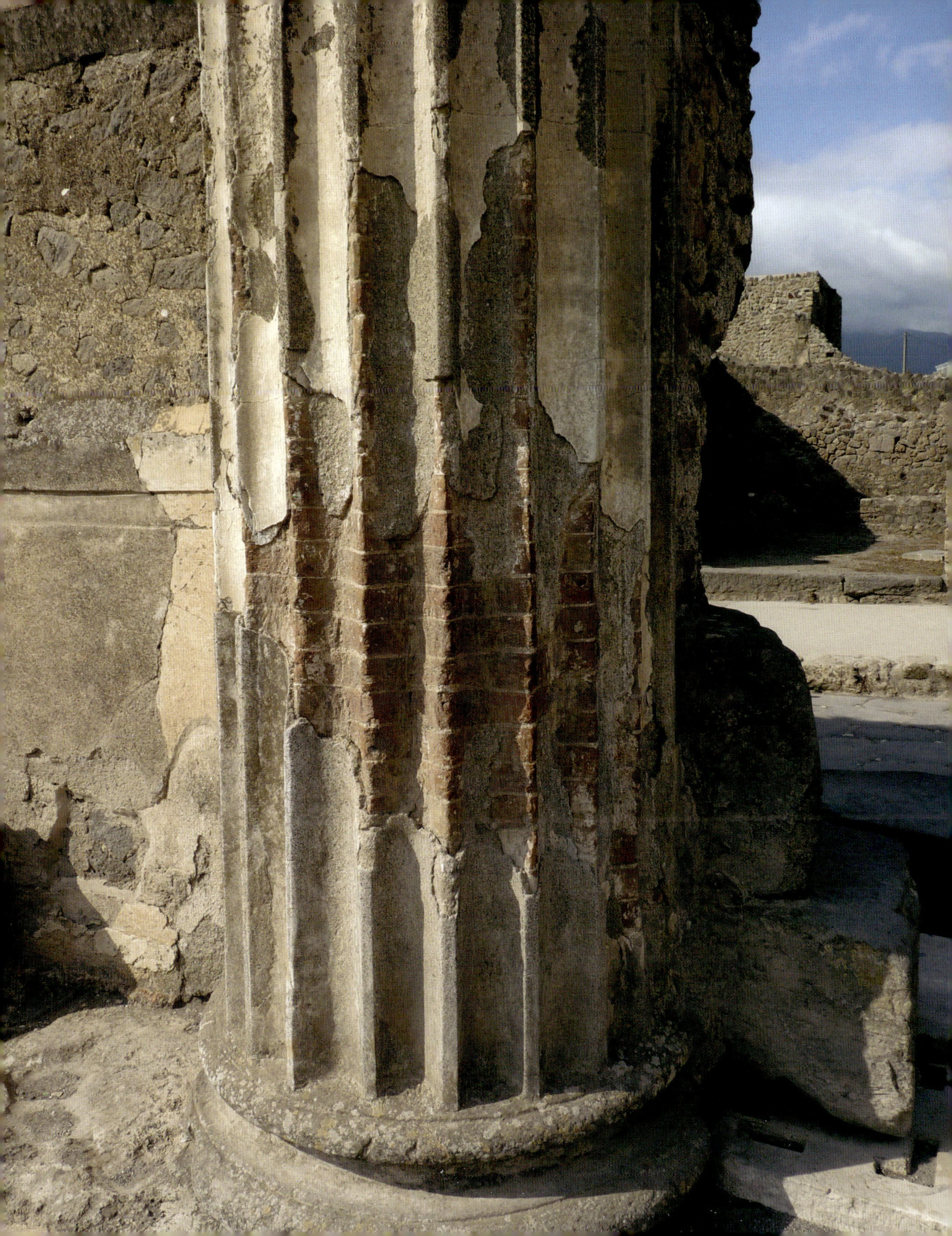

The Leptis Magna ruins of Windsor Great Park, taken from the Roman city in Libya in 1826. Repairs included reconstructing previously dismantled unsafe columns.

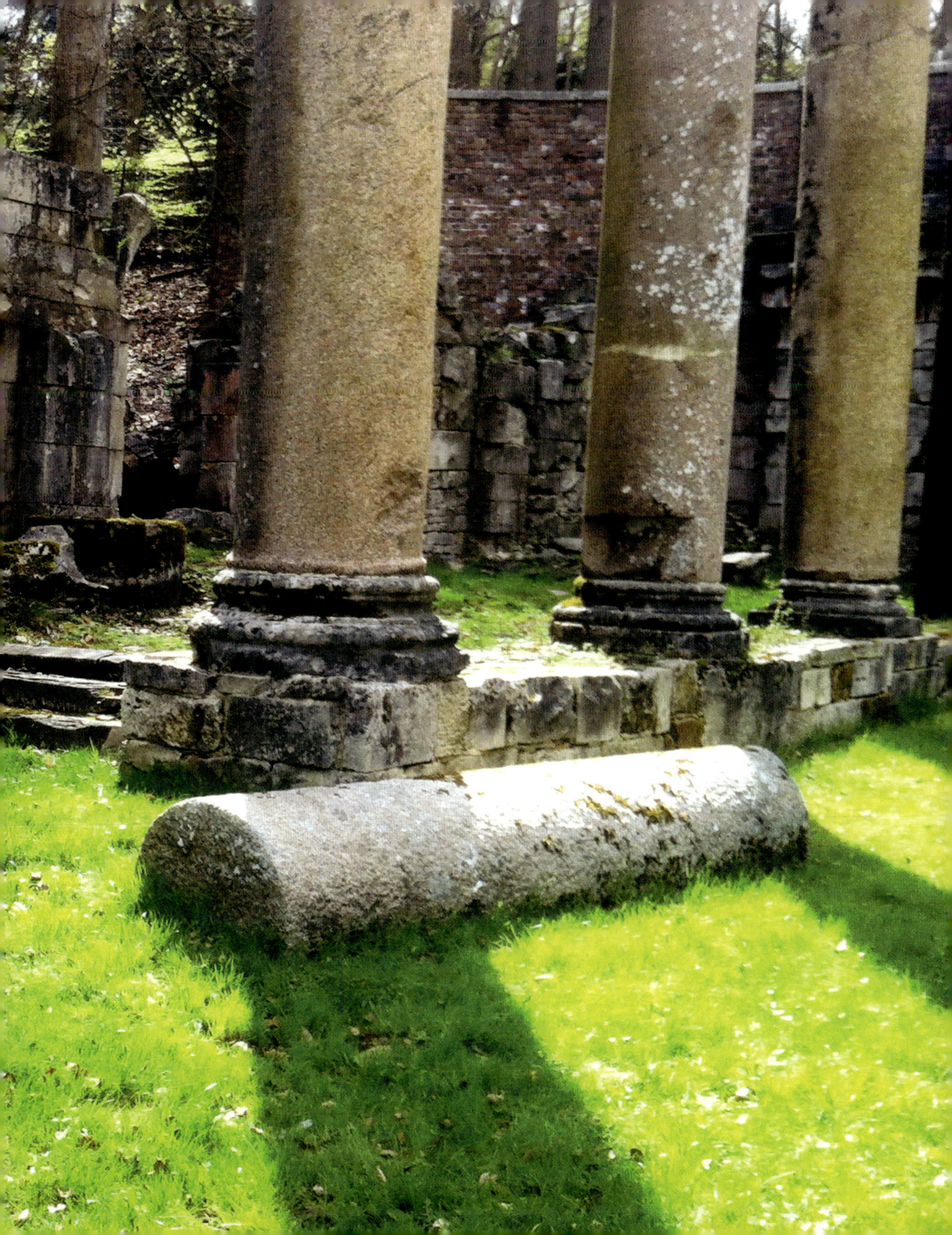

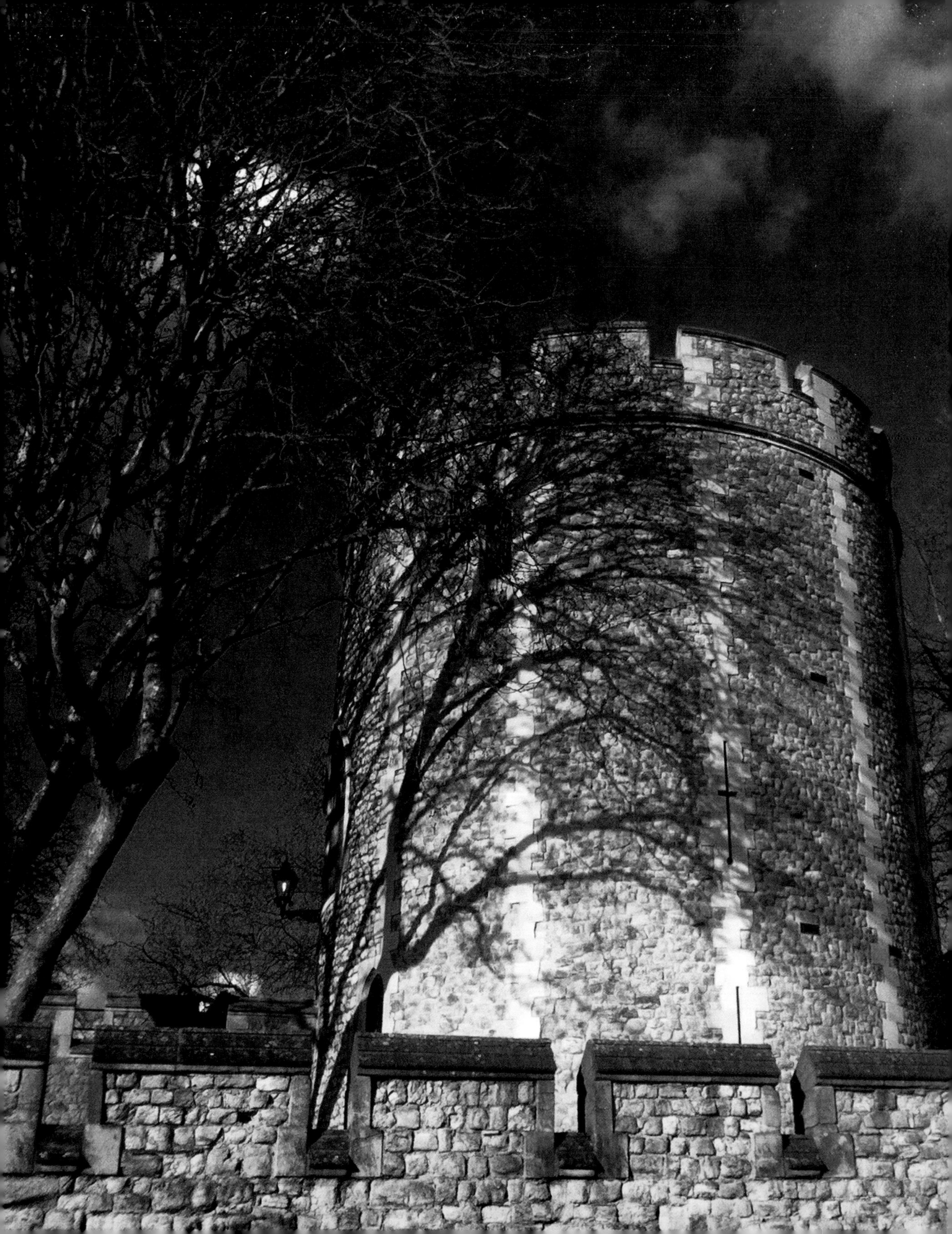

the building of a large castle in Caen, Normandy, for the future King of England using the local Caen limestone. They had previously applied their masonry expertise in southern Italy building, adapting, and strengthening numerous castles. Initially employed as mercenaries fighting against the saracen invaders, such was their success that the Pope appointed them Dukes of the southern Italian regions, and from here they went on to conquer Sicily, building and adapting older castles as they went. When they invaded Britain, they applied all their knowledge and did what they knew best: they commenced a massive programme of building castles. The important Norman castles in and around London are Rochester (strategically positioned where the river Medway meets the Thames), Windsor and the Tower of London.

A bishop from Normandy was appointed by William the Conqueror in 1070 to oversee the construction of the Tower of London. His name was Gundulf and, interestingly, the British Army's Corps of Royal Engineers regard him as their founding father. Rochester Castle, Rochester Cathedral, and several smaller forts were also built under the guidance of Gundulph at this time.

The stone used to construct Rochester Castle and the Tower of London was Kentish Ragstone, supplemented with Caen stone for the decorative carved arches and quoins. Nothing had previously been built in London on the awesome scale of the Tower.

Caen stone was also used to rebuild St Paul's Cathedral in 1087 after the previous building burnt down. The spire of this cathedral was 149m high, only 12m lower than the world's current highest church spire in Ulm, Germany.

The next few centuries saw further extensive use of Caen stone: Canterbury Cathedral from 1175, the first stone London Bridge from 1176, Westminster Abbey from 1252, and the Eleanor Crosses from 1290. The Norman masons' legacy is left around us not only in London and Britain, but much further afield in Normandy, northern Europe, Greece, Italy, and Sicily. The legacy also lives on in our language as evidenced by the fact that many of the words used to describe masonry are of french origin, for example mason, maul, banker (banquette),

FAR LEFT
Norman masonry at the Tower of London

ABOVE
Kentish Ragstone and Caen stone at the Tower of London

LEFT
The Roman London city wall in Kentish Ragstone with brick tile courses

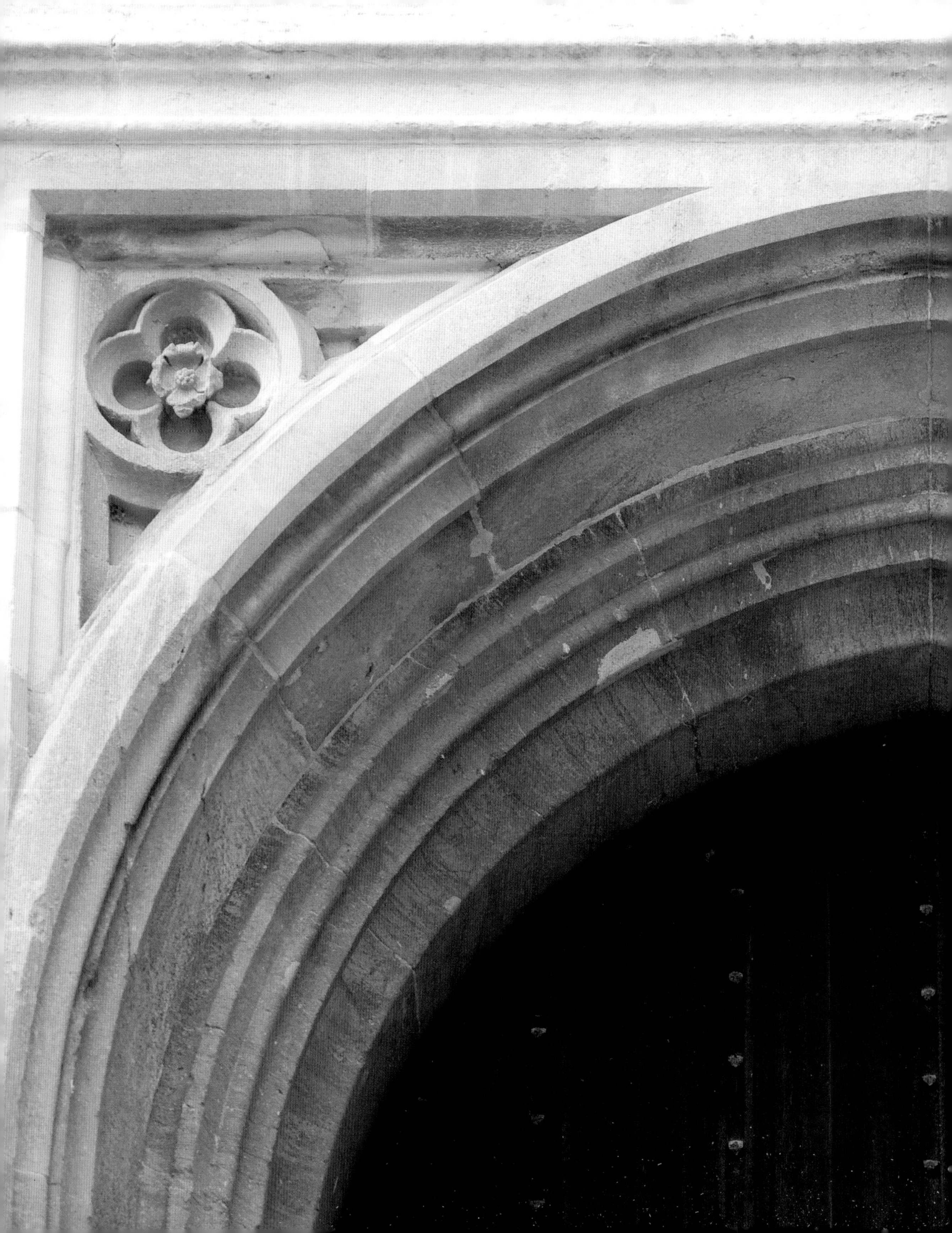

frenchman (a pointing tool), buttress, tracery, pinnacle, mullion, and transom.

The brittle difficult grey Kentish Ragstone and the contrasting workable light buff Caen stone were often used together and can be considered the signature stones of this period. What was left of the Roman city would have been plundered for the cheaply available Kentish Ragstone.

The use of Caen stone declined in the 14th century, and Reigate stone gradually became more widely used in its place. Reigate stone was difficult to quarry and expensive to transport, there being no suitable rivers nearby for barges to bring the stone to London. The stone had to be taken by ox-cart in small loads of little more than one cubic metre on a three day journey to Vauxhall, where it was loaded into barges and taken downstream to the city or across the river to Westminster. The cost of transporting this stone into London by ox-cart and barge would have been many multiples of the cost of the stone.

Another significant reason why Reigate stone should not have been an obvious choice was its poor durability. The early Guildhall was constructed using Reigate stone and only lasted 100 years before having to be demolished, and a contract at Eton College in 1453 instructed the masons not to use Reigate stone 'even in the foundations'.

The question why Reigate stone appears to have suddenly been used in preference to Caen stone might be answered by the fact that England was at war with the French from 1337 until 1453. The first phase of this war was fought in the English Channel, Caen and Calais from 1337 until 1360, and both sides used privateers to plunder each

ABOVE
Kentish Ragstone repairs at the Tower of London

LEFT
Kentish Ragstone and Bath stone detail at the Armenian Church, Kensington

FAR LEFT
Portland and Bath stone detail on the 15th Century façades of the Chapel at Eton College, Windsor

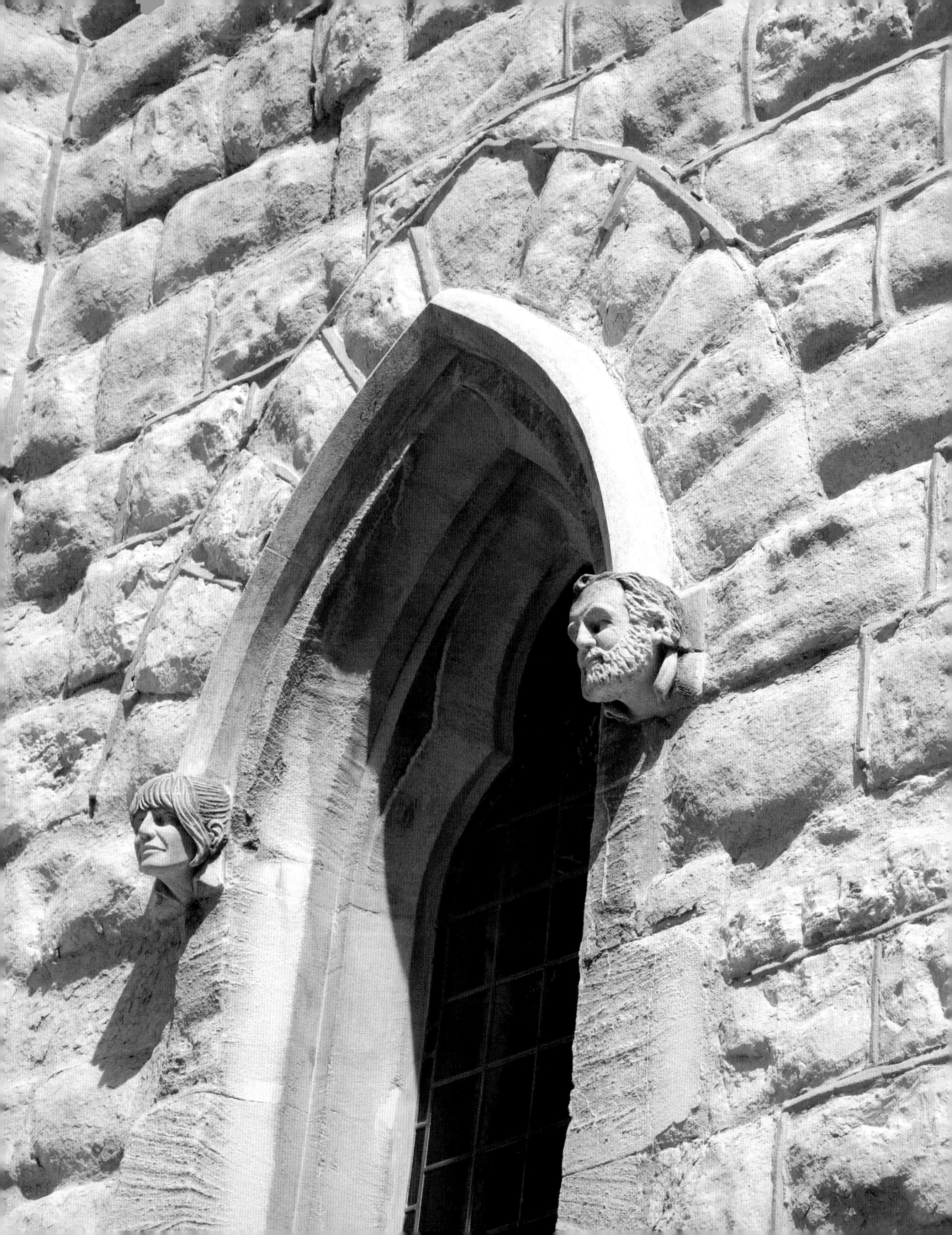

New label mould bosses in Bath stone on the cleaned and repaired Kentish Ragstone façades of the Armenian Church, Kensington

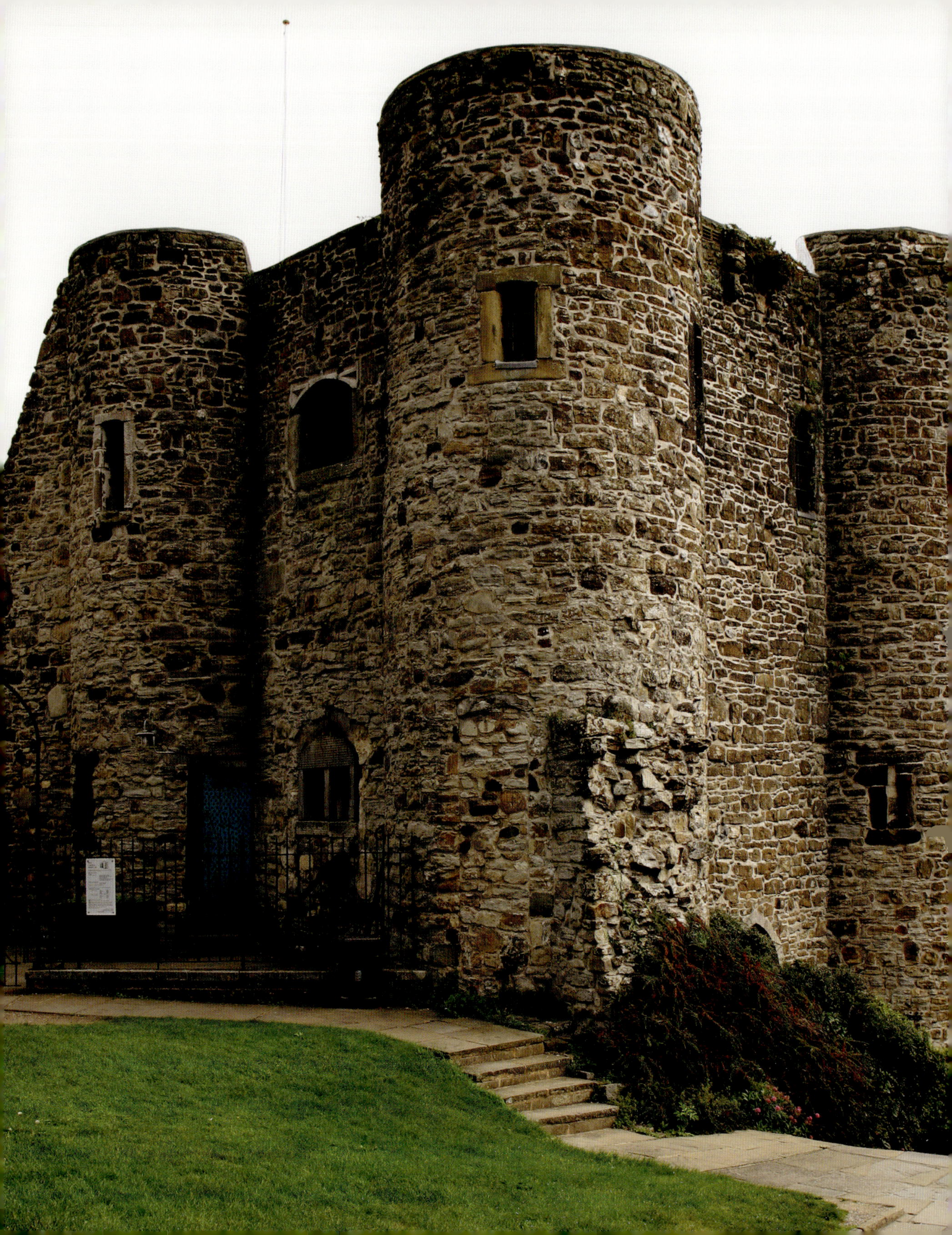

other's ships making the transportation of Caen stone impossible. The English Channel would have been a dangerous place until the Treaty of London was signed halting piracy in 1604.

During the period from the late 13th century, clay bricks were again used in London for more construction work at the Tower of London. These were manufactured and imported from Flanders. Bricks were first manufactured in England from approximately 1425 and early examples of buildings include Eton College in 1442, Crosby Hall in 1466, Eltham Palace in 1479, the Palace of Lambeth in 1490, Hampton Court in 1514, and St James' Palace in 1531.

THE EARLY MODERN PERIOD

From the middle of the 15th century there are few references to substantial buildings other than royal palaces. Not only does it seem that investment in grand structures was in decline, but the quality of workmanship, judged against the fragments of surviving masonry from this period, was particularly poor.

The reasons for this apparent decline are not clear but there would have been many contributing factors: the long war with France was lost in 1453; the War of the Roses from 1455 until 1485; and the Civil War from 1642 until 1651. Plagues had also ravaged England and Europe from the early 14th century until 1665, severely depleting populations and crippling economies.

London's narrow streets at this time were tightly packed with timber houses and shops. Fires were frequent. It has been written that fire alarms were raised so often during the night that many did not leave their beds until they could actually smell smoke. Eventually, a major incident occurred. The Great Fire of 1666 started in Pudding Lane on the 2nd September and raged for four days. It was the talk of Europe and destroyed more than 13,000 houses. Eighty-five percent of the city was lost together with 63 acres around Fleet Street.

This catastrophe brought much needed change, ushering in a new era.

AFTER THE GREAT FIRE OF 1666

Regulations were introduced which required building façades to be constructed using fireproof materials and for streets to be wider. A piped water supply was provided throughout the city.

England had steadily become wealthy during the 17th century. Trade and commerce had grown rapidly and industries such as glass making, brick making, iron and coal mining were flourishing.

LEFT
Ragstone façades of Ypres Tower at Rye Castle built in 1249

ABOVE
Bath stone indent repair on a 15th century façade

Sir Christopher Wren was appointed Royal Architect in 1669 and was made responsible for overseeing the reconstruction of London. Not only did this herald the new architectural style of restrained classicism sometimes referred to as English Baroque, but also a new material: Portland stone. Wren was not a fan of Reigate stone (describing its use as 'lamentable'), or Caen stone ('more beautiful than durable'), and Kentish Ragstone was not suitable for creating the clean and efficient moulded lines required by the new architecture. Portland stone was selected for the reconstruction of St Paul's Cathedral in 1671, and it quickly became London's stone of choice for all important buildings.

Portland stone had been used in London before this time. Inigo Jones had used Portland stone for dressings on the façades of the Whitehall Palace (now the Banqueting House) in 1622, and the Norman Caen stone cathedral had been repaired using Portland stone between 1633 and 1642. Some accounts suggest Portland stone had been used even earlier in a short two year period in the mid 14th century at the Tower of London and on the London Bridge (though, if true, this was an isolated occurrence and seemingly out of context).

There are many reasons why Portland stone was selected: its attractive appearance, its good durability and its convenient quarry access to natural harbours on the Isle of Portland. There were political reasons too. The Crown owned the quarries, and Wren himself was MP for Weymouth, the nearest town to the Isle of Portland.

Six million tons of Portland stone were used to reconstruct London in the years following the Great Fire. This equates to more than £15,000 million at current rates excluding transport and installation costs (and VAT!). The demand for labour in the Portland quarries would have been tremendous. The Crown's response was to establish prisons there, setting the convicts to work in the quarries.

THE POST-INDUSTRIAL PERIOD

The 18th century use of Bath stone in constructing the wealthy cities of Bath and Bristol established a successful stone industry in the West Country. However it was not economically viable to transport stone from Bath to London by barge until the Kennet and Avon canal was finally extended to the Thames in 1810. Bath stone is easier and cheaper to carve than Portland stone, though not as durable in polluted cities. It did however provide choice as a cheaper alternative.

ABOVE
Portland stone façades of Wren's Royal Hospital for Seamen at Greenwich built between 1696 and 1712

RIGHT
Portland stone façades of the 18th century Darnley Mausoleum

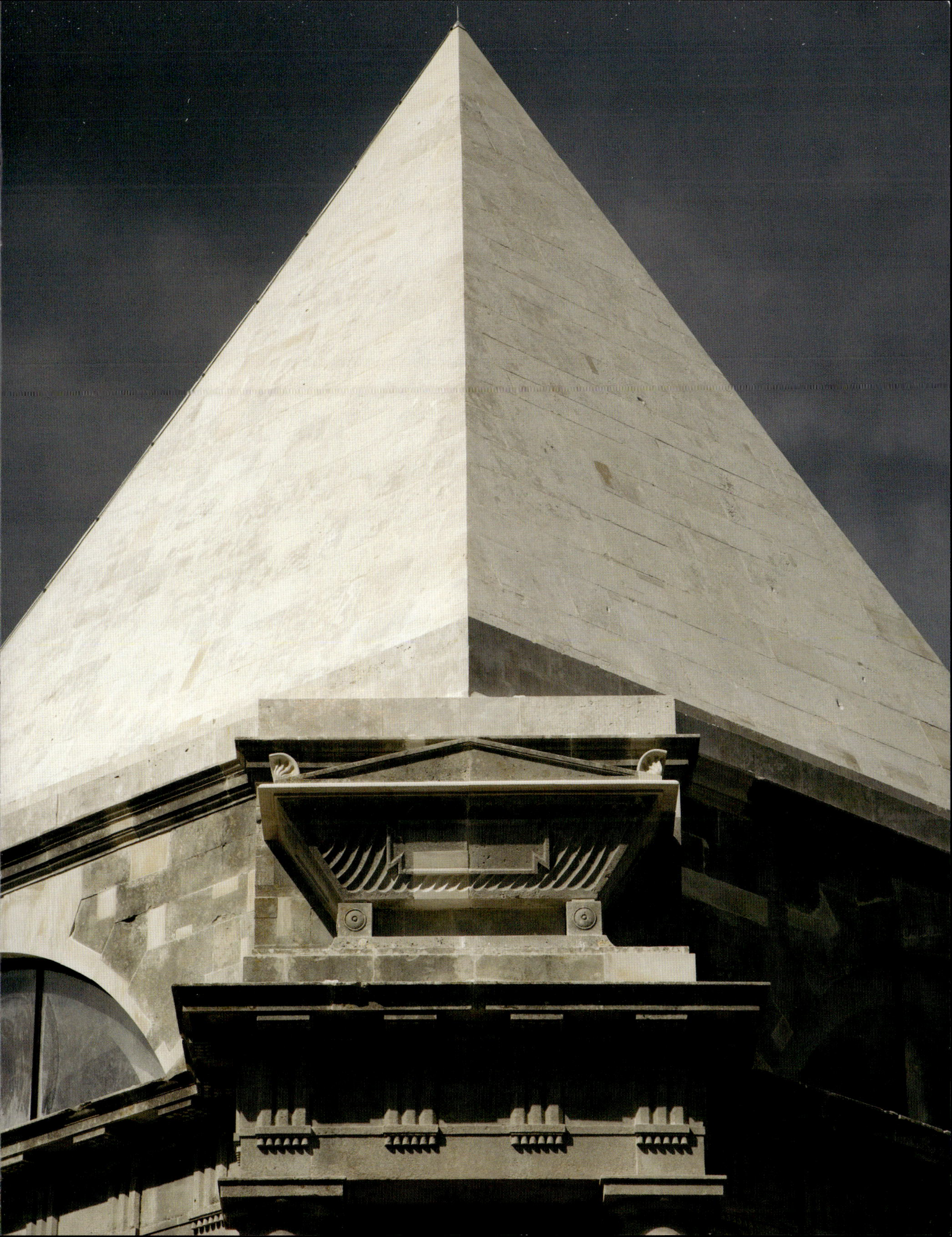

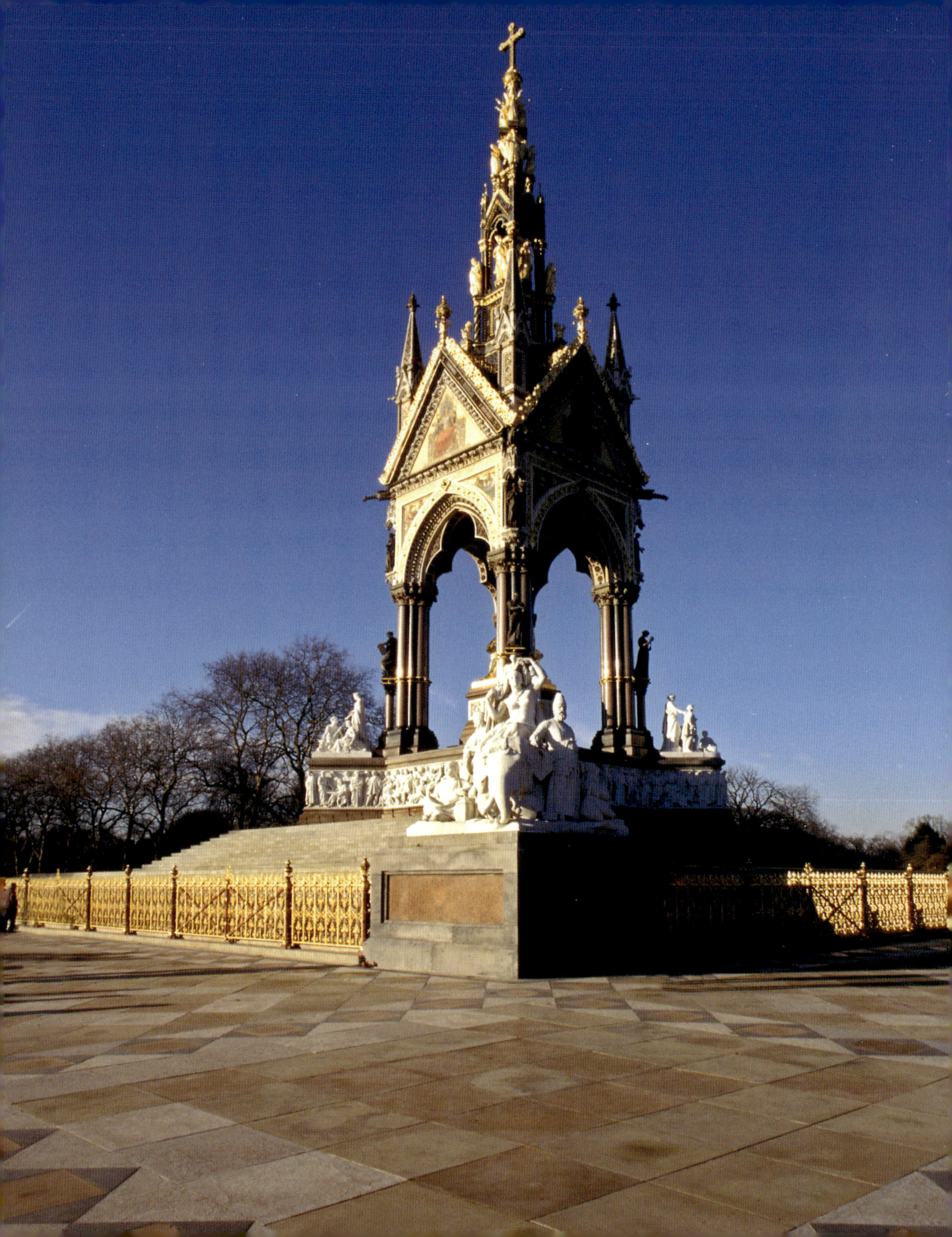

Soon after the establishment of the river and canal system for transporting freight, steam emerged as the new power. In 1835 the Great Western Railway Company opened the London to Bristol line (with a loop to Bath) further reducing the cost of transporting Bath stone into London. The Northern and Midland lines were extended to London in 1846 which, perhaps not coincidentally, passed through stone quarrying areas such as Ancaster, Mansfield and Blaxter facilitating the introduction of these stones into London for particular use in railway stations and bridges. An excellent example is the St Pancras Hotel and Station which were designed by George Gilbert Scott using materials brought to site by railway. Several other railway stations, embankments, and bridges were built using these stones.

The construction of the Albert Memorial in 1872 demonstrates the range of stones used and available in London by this time. Obviously the usual budget constraints would not have applied but it does serve to illustrate the materials then popular in London: granite from Scotland and Cornwall, Portland stone, York stone, Red Mansfield stone, slate from Ireland, and marble from Carrara, Italy.

City atmospheres had become severely polluted from the burning of coal and this was taking its toll on the stone façades. Even Portland stone ceases to be durable if harmful pollution is allowed to accumulate on its surface. The pollution in London was severe.

New materials resistant to pollution damage were sought, and this requirement was satisfied by the introduction of terra cotta as an early form of commercially viable artificial stone. Its popularity increased in London from the late 19th century into the early 20th century.

Terra cotta is made by pressing clay into plaster moulds, drying, and kiln-firing the blocks. Economies of scale are achieved through the repetitive use of moulds. A consequent characteristic of all terra cotta façades is the repetition of details and mouldings. Manufacturers sought to maximise economies of scale by standardising components so an architect could select cornices, string courses and other details from catalogues. This worked to an extent with developers such as Joe Lyons in the early 20th century selecting glazed terra cotta to construct the façades of his new Savoy, Strand Palace and Regent Palace Hotels in successive years.

Terra cotta was used as an economic and durable alternative to natural stone until the outbreak of the First World War in 1914. The units were created as three dimensional hollow blocks and generally detailed in a similar manner to stone. However, after the war, styles of architecture changed and hollow blocks gave way to more

LEFT
Portland, York, Red Mansfield, slate, Cornish granite, Scottish granite and Carrara marble at the Albert Memorial, completed in 1872

ABOVE
Bath stone being transported by barge on the Kennet and Avon canal which opened in 1810 allowing the stone to be economically used in London, Windsor and Oxford

economic two dimensional tiles which were particularly suitable for use on art deco façades such as the many cinemas constructed in the 1920's and 1930's.

Bath stone was still being brought into London by rail and Portland stone transported by barge until the outbreak of the Second World War. It was not commercially viable to transport stone long distances by road until the construction of trunk roads and motorways from the mid-20th century.

THE INDUSTRY TODAY

Exemplory skills thrive in the UK but are increasingly threatened by the developing tendancy to import stone worked abroad regardless of the environmental cost of shipping heavy materials around the world. In the 19th century, Britain was the world's largest exporter of granite. Now, what little stone flows from Cornwall and Scotland rarely gets beyond the UK's borders. It is difficult to understand how it can cost less to quarry and import granite from China than from Cornwall, but it does, and this is the new reality.

Demands to provide commercial floor space as cheaply as possible drive the market away from stone towards glass and steel, however we are fortunate that the UK has a world class industry that provides the skills and stone for the maintenance and repair of our historically important landmark buildings. This industry too can provide the expertise and materials for discerning developers wishing to create the landmark buildings of the future.

ABOVE
Terra cotta façades forming the south portico under construction at the Royal Albert Hall

RIGHT
Red Mansfield stone on the late 19th century façades of the London Hippodrome

LEFT
Bath stone façade in Fleet Street

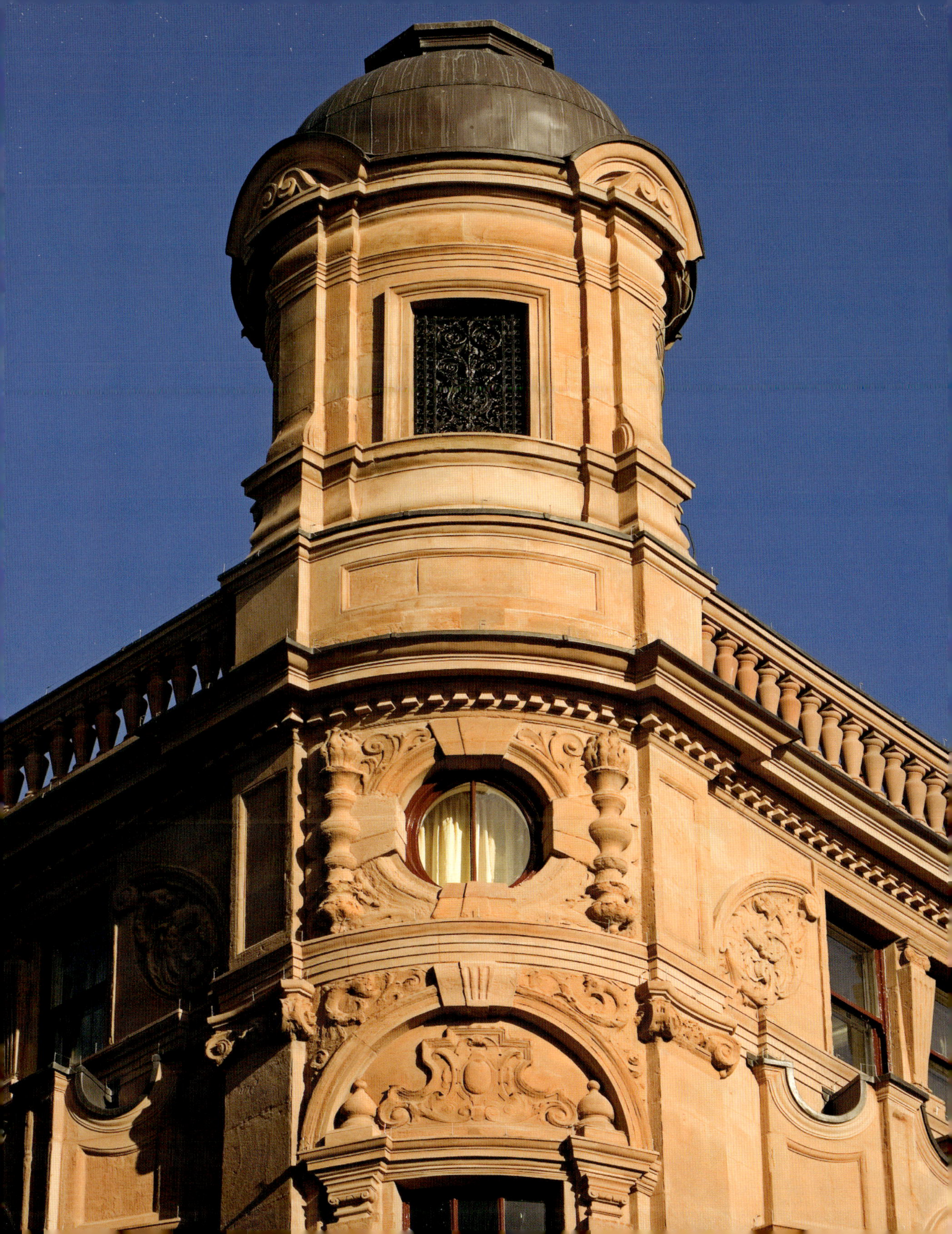

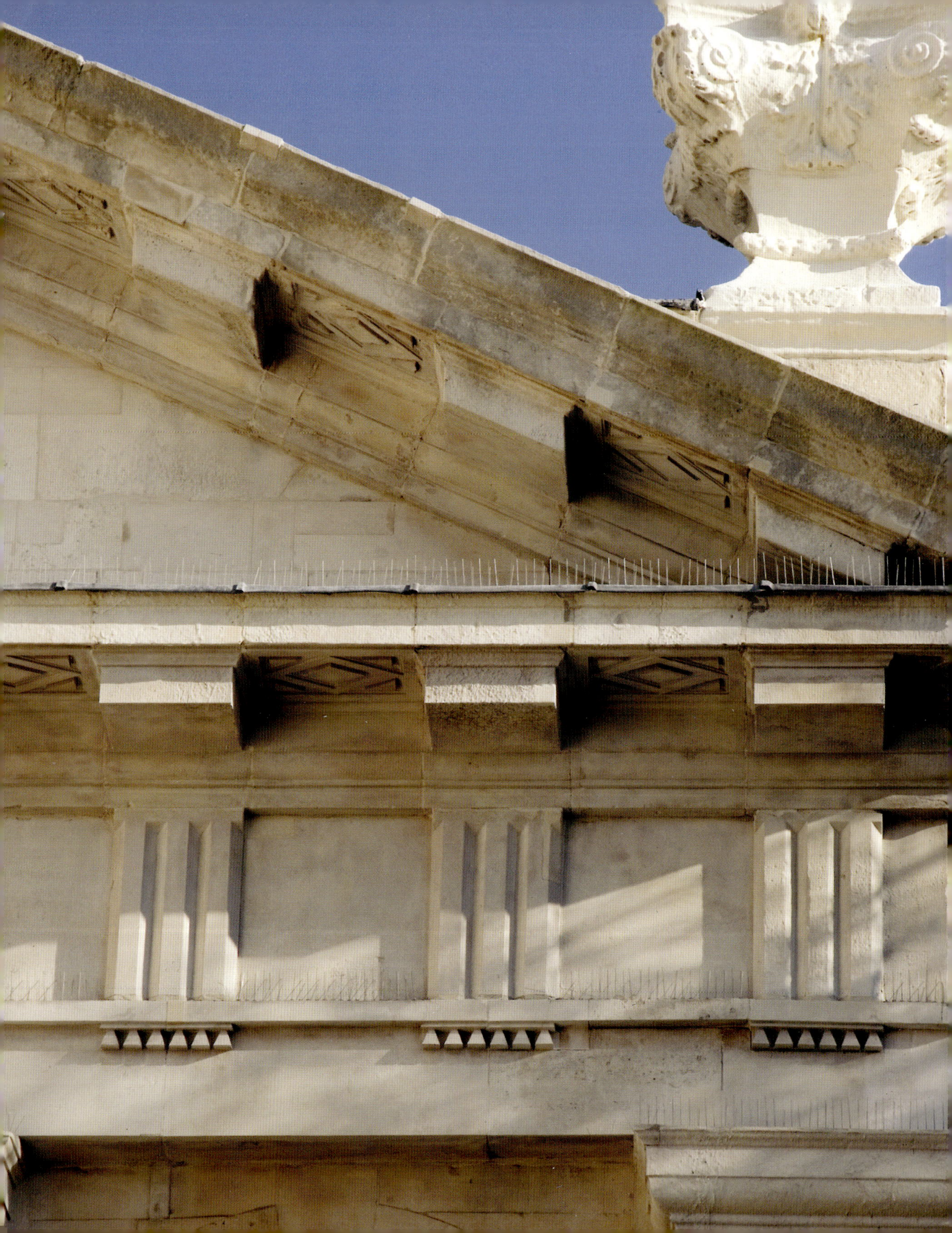

A STONEMASON'S HISTORY OF LONDON

FAR LEFT
Portland stone on the 18th century façades of St Alphege Church, Greenwich

ABOVE
Inspecting a block of Portland stone in the Perryfield Quarry

LEFT
Bath stone detail, St Matthias, Richmond

GLOSSARY

A

ABACUS - The flat slab on top of a capital

ABRASIVE HARDNESS - The wearing quality of stone

ACANTHUS - A plant with leaves on which the Corinthian and Composite capital are based

ACID ROCK - Igneous rock containing more than 66% silica

ALABASTER - Fine-grained compact pale-coloured variety of gypsum

ALCOVE - A recess in a wall

ALTAR - A slab or table used for religious ceremonies

AMBERLEY - Chalk quarried in Sussex

AMBO - A lectern in early churches

AMMONITE - Spiral fossil

ANGLE-TOOLED - Dressed with diagonal tool marks

ANTEFIXAE - Small decorative blocks on a parapet

APEX STONE - Top stone of a gable, spire or pediment

APRON - A panel below a window cill

APSE - The east end of a chancel or nave

ARCADE - A range of arches on piers or columns

ARCHITRAVE - (1) The lowest of three parts of an entablature (2) The moulding round a door or window

ARCHIVOLT - An architrave moulding following the contours of an arch

ARENACEOUS - Composed of sandy grains

ARGILLACEOUS - Containing clay minerals

ARRIS - External edge produced by the meeting of two planes

ARTIFICAL STONE - Cast cementitious material containing crushed stone

ASHLAR - Walling of plain blocks of stone

ASTRAGAL - A small moulding semi circular in section

ATRIUM - A covered inner courtyard

B

BALUSTER - Vertical member of a balustrade

BALUSTRADE - Barrier formed by a series of balusters surmounted by a coping

BAND COURSE - Horizontal feature course either flush, projecting or recessed

BANKER - Heavy bench made of timber, blocks of stone, etc

BANKER MASON - Mason who works stone on a banker

BAPTISTRY - Part of a church containing the font

BARREL VAULT - A simple vault unbroken by cross vaults

BAS-RELIEF - Decoration carved to project from the background

BASALT - Fine-grained igneous volcanic rock

BASE BED - The lower usable quarry bed

BASE COURSE - Lowest course of a wall

BASIC ROCK - Igneous rock containing more than 45% and less than 52% silica

BATTED - A regular pattern of fluted cuts in the stone face

BATTLEMENT - Finishing of a parapet wall with alternative openings and projections

BEAD - A small moulding round in section

BEAD-AND-REEL - An enriched moulding with alternative spheres and cylinders

BEARING - That part of the stonework that takes the load

BED - (1) Layer of sedimentary rock (2) Surface on which a masonry unit rests

BED JOINT - Horizontal mortar joint

BED-MOULDING - The lower section of a cornice

BIRDSMOUTH - Recess formed along the junction of the external mitred joint of masonry cladding by blunting the sharp arris of each stone

BLIND TRACERY - tracery used as a decoration feature on a plain wall

BLOCK STONE - Stone roughly squared at the quarry

BLOCKING COURSE - Plain course above a cornice acting as a counter balance

BOASTED - Stone finish produced by a mallet and boaster to form bands of parallel lines

BOLLARD - Short strong post

BOND - Arrangement of masonry units

BONDER - Stone used to tie through the thickness of a wall

BRACKET - A supporting piece of stone to carry a projecting weight

BROACH - A half pyramid connecting the angle of a square tower with the face of an octagonal spire

BUCKET HANDLE JOINT - Morter joint tooled to a concave finish

BUTTRESS - A mass of masonry to provide lateral support

C

CABLE-MOULDING - A moulding imitating a twisted rope

CABOCHON - Diamond shaped inserts in a tile floor

CALCAREOUS - Containing significant amount of calcium carbonate

CALCITE - Crystalline form of calcium carbonate

CANOPY - A hood over a door, window, niche or tomb

CANTILEVER - Horizontal projection fixed at one end and counter balanced

CAPITAL - The head of a column

CARTOUCHE - Panel in the form of a sheet of paper with curling edges

CARYATID - A column carved in the form of a human figure

CAST STONE - Artificial stone

CASTELLATED - Wall with battlement on top

CATHODIC PROTECTION - A system whereby the embedded steelwork being protected against corosion is made to act as a cathode

CAVETTO - A hollow moulding a quarter of a circle in section

CENTERING - Temporary wooden structure on which arch is built

CHAMFER - Flat surface formed by planing off sharp angle at intersection of two surfaces

CHANCEL - The part of the east end of a church in which the alter is placed

CHOIR - That part of a church between the nave and chancel

CINQUEFOIL - A five lobed foil window

CLERESTORY - The upper levels of the main walls of the nave above the roof of the aisle

CLOISTERS - A vaulted, open-sided passage facing onto a quadrangle

CLUNCH - Chalk used for building

COADE STONE - A type of terracotta used since the 1770s and for much of the 19th century

COBBLE WALL - Wall built with beach cobbles

COFFER - A sunk panel in a ceiling, dome or vault

COLONNADE - A row of columns carrying an entablature or arches

COLUMN - Vertical member either free-standing or supporting a lintel

CONSOLE - An ornamental bracket

COPING - Overhanging protective covering for the top of a wall

CORBEL - Cantilevered stone of one or more courses that projects to form a bearing

CORE - Internal filling of rubble in a wall

CORNERSTONE - Quoin

CORNICE - A horizontal moulded projection usually composed of three sections the cymatium, the corona or drip section, and the bed moulding

CORONA - The drip section of a cornice

COURSE - Layer of masonry units of uniform height

COVING - The concave moulding at the junction of a wall and ceiling

CREASING COURSE - Course of tiles or slates laid in a bed joint projecting to form a drip

CRENELLATION - A battlement

CROCKET - Decorative carved leaf shapes projecting at regular intervals from the angles of spires, pinnacles, canopies, gables etc in Gothic architecture

CROSSING - The space at the intersection of the nave, chancel and transepts of a church

CROWN - The highest part of the curve of an arch or vault

CRYPT - A vault beneath the main floor

CUPOLA - A dome on a circular or polygonal base, above a roof or turret

CURTAIL STEP - The lowest step in a staircase, with a curved end that projects beyond the newel

CUSP - Projecting points formed at the meeting of the foils in Gothic tracery

CUTWATER - The wedge-shaped pier of a bridge, constructed to break the current of water

CYMA RECTA - An ogee moulding, concave above and convex below

CYMATIUM - The upper section of a cornice

D

DADO RAIL - A rail at approximately waist high

DENTIL - A series of small square blocks in the decorative part of a cornice

DIAPER WORK - Decoration composed of small repeated patterns

DIE - Intermediate solid pier within a balustrade

DOME - Hemispherical vault, circular on plan

DOWEL - Short piece of metal or slate sunk into adjacent faces to align and/or prevent movement

DRAFTED MARGIN - Tooled margin worked on the edge of a rough-faced stone

DRAG - Tool with teeth for finishing the flat surface of soft stone

DRAGGED - Having irregularities on the exposed surfaces of soft stones worked off by the use of a drag

GLOSSARY

DRESSED - Any kind of worked finish

DRESSINGS - General term for decorative masonry features

DRIP - Projection below a horizontal surface to prevent rainwater flowing back to a wall

DRUM - (1) A vertical wall supporting a dome or a cupola (2) Individual stone in the shaft of a column

DRY STONE WALL - Stone wall constructed without mortar

DUTCH GABLE - A gable curved on elevation

E

EFFLORESCENCE - Soluble salts brought to the surface of masonry by the evaporation of moisture

EGG-AND-DART - An ovolo moulding decorated with a pattern based on alternate ovals and arrowheads

EMBRASURE - A small opening in the parapet of a fortified building

ENTABLATURE - A cornice consisting of architrave,

ENTASIS - The slight curve on columns to accentuate their height

EXFOLIATION - Scaling of stone caused by weathering

EXTRADOS - Convex surface of an arch

F

FACE BEDDED - Stone laid with the bed plane running vertically and parallel to the face of a wall

FAIR END - A clean cut end of a stone

FALSEWORK - Temporary support used in construction

FASCIA - A plain, horizontal band

FERRAMENTA - Metal bars used to support leaded windows

FESTOON - A curved garland of fruit and flowers

FETTLING - Hand dressing the joints of a paving slab

FINE AXED FINISH - Axed finish to granite

FINIAL - A ornament at the top of a gable or pinnacle

FLAGSTONE - A paving stone made from sedimentary rock easily split into flat slabs

FLAME TEXTURED - Rough face on surface of granite achieved by spalling the face with a high-temperature burner

FLEXURAL STRENGTH - The property of a material to resist flexing

FLINT - Cryptocrystalline silica originating as nodules in layers in chalk

FLUSH JOINT - Mortar joint the surface of which is level with the face of adjacent masonry units

FLUTING - Shallow, concave grooves running vertically on the shaft of a column, pilaster or other surface

FLUTED - Surface worked into a regular series of concave grooves

FLYING BUTTRESS - An arch transmitting lateral thrust from the upper part of a wall to an outer buttress

FLYING STAIRCASE - A staircase where the steps are cantilevered

FOIL - A lobe or leaf-shaped curve within a tracery window arch

FOOTING - Stepped courses of masonry in foundation

FREE LIME - Lime in a mortar which remains as calcium hydroxide rather than carbonating

FREESTONE - Stone workable in any direction

FRIEZE - The middle division of an entablature

G

GABLE - The triangular upper portion of a wall at the end of a pitched roof

GABLET - A decorative motif in the form of a small gable

GALLETING - Small pieces of stone or flint in the face of joints

GARGOYLE - Projecting masonry feature possibly forming a water spout

GOTHIC - Term for a period of architecture based on the pointed arch

GOTHIC COPING - Coping with roll top, weathered and moulded

GRANITE - Acid, igneous, plutonic rock

GREENSAND - Sandstone coloured with glauconite

GRITSTONE - Sandstone with angular, usually coarse grains

GROIN - The intersection between two vaults

GROUT - Liquid mortar fed into a wall to fill voids

GUTTAE - Small projections carved below the tenia under each triglyph on Doric architrave

H

HANDED - Elements that mirror each other

HARDNESS - Resistance to abrasion depending on cohesion of grains and hardness of constituents

HAWK - Tool for holding mortar ready for application by trowel

HEAD - Lintel

HEAD TREE - Temporary support below a lintel or arch

HERRINGBONE WORK - Work in which the units are laid diagonally. Alternate courses lie in opposite directions

HOLLOW BEDDING - Setting of blocks with mortar at the ends only, the centre section being left hollow to prevent breakage in case of settlement

HONED - Having a dull polish or a matt surface

HOOD-MOULD - Moulded canopy over a door or window opening providing weather protection

HYDRATED LIME - Product obtained by slaking quicklime with water

HYDRAULIC LIME - Quicklime able to set in the presence of water. Hydraulic lime is said to be emminently, moderately or feebly hydraulic

I

IGNEOUS - Rock formed by the cooling and consolidation of magma

IMPOST - Masonry unit from which an arch springs

INDENTING - Piecing-in

INTERMEDIATE ROCK - Igneous rock containing more than 52% and less than 66% silica

INTRADOS - Concave surface of an arch of its line on the face of an arch

J

JAMB - Stone forming the side of an aperture

JOGGLED - Arch in which adjacent voussoirs are interlocked by means of visible rebates or steps

JOINTING - Finishing mortar joint as building work proceeds

JUMPER - A large bonder stone rising through two or more courses of a wall

K

KERF - Groove made by a saw

KEY PATTERN - A geometrical ornament of horizontal and vertical straight lines repeating to form a band

KEYSTONE - Central locking voussoir

KNAPPED FLINT - Flint nodules split across and used in walls with the split face showing

KNEELER - Internediate stone with sloping top and level bed in a ramped coping

L

LABEL-MOULD - Projecting moulding providing weather protection over a door or window

LABEL-STOP - An ornamental boss at the terminations a hood-mould

LANCET - An arch where the radius is larger than the span

LANCET WINDOW - A pointed-arched window

LANDING - A flat slab forming the top of a staircase

LEACHING - Removal of mineral salts by percolating water

LEDGER - A large flat stone capable of spanning a space and carrying a load

LIME PUTTY - Slaked lime produced by mixing hydrated lime and water

LIMESTONE - Sedimentary rock consisting predominantly of calcium carbonate

LINTEL, LINTOL - Load bearing stone spanning across an opening

LOAM - Mixture of clay and sand (possibly reinforced with straw)

M

MAGNESIAN - Containing appreciable amount of dolomite

MARBLE - Metamorphosed limestone

MASON'S MITRE - Birdsmouth joint on an arris

MERLON - The sollid portions at a battlemented parapet between embrasures

METAMORPHIC ROCK - Rock recrystallised from pre-existing solid rock masses by the action of heat, and/or pressure

METOPE - The square space between two triglyphs in the frieze of a Doric order

MODILLION - A bracket supporting a classical cornice

MONOLITH - A large single stone

MOULDINGS - The contours given to projecting members

MULLION - Vertical divider of window

MUTULE - The projecting square block above the triglyph and on the soffit of a Doric cornice

N

NATURAL BEDDING - Laying a stone with the bed plane horizontal as quarried

NATURAL FACE - As from the quarry

NAVE - The main body of a church

NECKING - (1) A narrow moulding round the bottom of a capital (2) The narrowest part of a baluster

NICHE - A recess in a wall

GLOSSARY

O

OBELISK - A tall tapering shaft of stone of square section

OCULUS - A round window

OGEE CURVE - A double-curved line

OOLITE - Small, spherical grains of calcium carbonate

ORDER - In classical architecture, a column with base shaft, capital, and entablature, Doric, Tuscan, Ionic, Corinthian or composite

ORIEL - Bay window off the ground

P

PARGING - The rendered lining to a chimney

PATERA - A small flat individual repetitive carved floral ornament

PEDESTAL - Base to column or statue

PEDIMENT - Triangular gable

PENDANT - A ceiling boss elongated so that it hands down

PENDENTIVE - The vaulted in-filling at the angles where a polygonal structure charges to circular

PERPEND - Vertical joint appearing in the face of a wall

PIER - Solid masonry support

PILASTER - Part of a column or shallow pier projecting from a wall

PITCHED - Surface produced by a pitching tool to resemble the natural rock face

PLINTH - Projecting base of wall or column

POINTING - Filling to mortar joint

POROSITY - Ratio of volume of pore space to volume of rock given as percentage

PORTE-COCHERE - A porch large enough for wheeled vehicles to pass through

PORTICO - The entrance and centrepiece of the fa√ßade

POULTICING - Method of drawing soluble salts or stains out of stone by applying an absorbent paste

POZZOLANA, POZZOLANIC - Originally a volcanic earth from Pozzuoli (Italy)

Q

QUARRY BLOCK - Roughly squared stone from the quarry

QUARRY FACED - Natural face

QUARRY SAP - Moisture in the stone when in the ground

QUICKLIME - Product obtained when chalk or limestone is heated to drive off the carbon dioxide

QUOIN - Featured external corner

R

RAG - Hard coarse rubble stone eg Kentish Ragstone

RAMP-AND-TWIST - A stone curved in two dimensions

RANDOM RUBBLE WALL - Rubble wall laid uncoursed

REBATE - Continuous sinking

RECONSTRUCTED/RECONSTITUTED STONE - Cast stone

REEDED - Surface consisting of small convex ridges

RELIEVING - An arch in a solid wall to divert some of the load away from the lintel below

REREDOS - A wall or screen behind an altar respond half-pier bonded to a wall, carrying one end of an arch

RETICULATED - Network of bands worked into the stoneface to a gauge and the face 'picked' with a fine mallet-headed point

RETURN - Visible surface at a change of direction

RETURNED END - Termination of stone worked to form external angle

REVEAL - Return at each side of an opening

RIB - A projecting band on a ceiling or vault, seperating the webs of a groined wall

RIBBON POINTING - Pointing following the shape of rubble units and finished slightly proud of the surface. Not regarded as good practice

RISER - Vertical surface of a step

RIVEN - Surface produced by splitting a stone along the cleavage planes

ROLL MOULDING - Moulding of more than semi-circular section

ROMAN CEMENT - Natural cement produced from 1796 by firing a calcareous clay stucco in the 19th century

ROSE WINDOW - A circular window with foils or patterned tracery

ROSETTE - A rose-shaped patera

ROTUNDA - A room circular in plan and usually domed

RUBBED - Smooth finish obtained by rubbing stone with abrasive

RUBBLE - Walling stone of irregular shape and size

RUBBLE WALL - Wall built with stones of irregular shape and size

RUSTICATED - Stone having a sunk, dressed margin to provide recessed joints

208

S

SADDLE BARS - In window glazing, the small iron bars to which the lead panels are tied

SADDLEBACK COPING - Coping weathered both ways from centre of section

SANDSTONE - Sedimentary rock composed of sand grains naturally cemented together

SCABBLED - Roughly faced with pick or hammer

SCAGLIOLA - Cement or plaster and marble chips or colouring matter to imitate marble. Popular in the 17th and 18th centuries

SCANT - Slab of stone sawn on two sides

SEDIMENTARY ROCK - Rock formed by deposition, usually in water, of particles of inorganic or organic origin

SETT - Stone roughly squared for roads and paving

SHAFT - The trunk of a column between the base and capital

SHAKE - Fractures that occur across the cleavage plane

SINKING - A recess

SLAKED LIME - Product formed when quicklime is slaked with water

SLAKING - Adding water to quicklime to form putty or dry hydrate

SLIP - Narrow piece of stone inserted between large blocks

SPRINGER - (1) Kneeler at the base of a ramped coping (2) The lowest stone of an arch

SPRINGING LINE - Level from which an arch springs

SQUARE - Tool for determining and testing right angles

STITCH - Reinforcement introduced into a wall to bond fractures

STOOLING - Raising of a weathered surface, to form a level seating

STOPPED END - Termination of moulding by returning the moulding into the wallface

STRING COURSE - Horizontal moulded projection incorporating a drip

T

TESSERAE - Small cubes of marble or stone used in mosaic

THROAT - Groove in an under-surface, designed to prevent water running back across it

THROUGH STONE - Bonder stone that extends through the thickness of a wall

TOOLED - having a dressed finish showing tool marks

TOOTHING - Masonry units left projecting or recessed in alternate courses to bond with future work

TRACERY - The ornamental intersecting work in a window, screen, or panel, or used decoratively in blank arches and vaults

TRANSEPT - The transverse arms of a cross-shaped church

TRANSOM - Horizontal member of a window

TREAD - Horizontal surface of a step

TRIGLYPHS - Projecting blocks with vertical grooves between the metopes in a Doric frieze

TYMPANUM - The space contained within a triangular pediment

U

UNDERCROFT - a vaulted room below an upper room

V

VAULT - Arched masonry roof with curved soffit

VEIN - A sheet of mineral present in rock joint or fault

VENT - Naturally occurring fault in a stone often a source of weakness

VERMICULATED - Having a dressed finish in the form of irregularly shaped sinkings

VOLUTE - A spiral scroll on an Ionic capital

VOUSSOIR - Wedge-shaped masonry unit in an arch

W

WATER TABLING - Weathered upper surface of any projection of masonry

WEATHERED - Sloped to throw off rain water

X

Y

YORK STONE - A generic term for sandstones from Yorkshire

Z

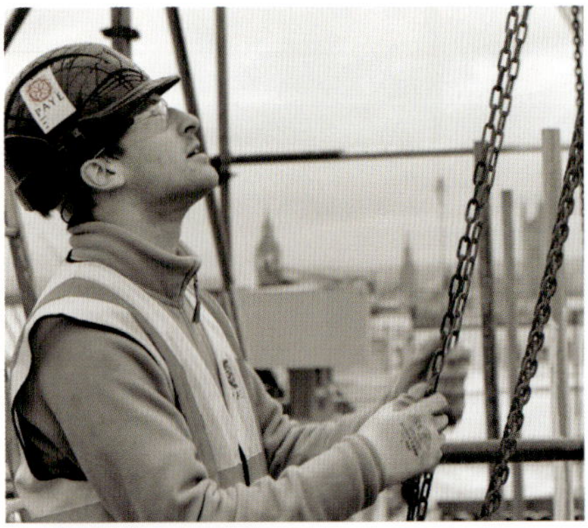
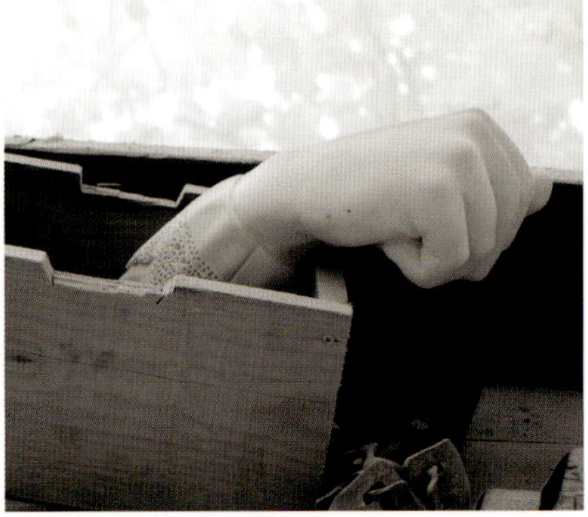
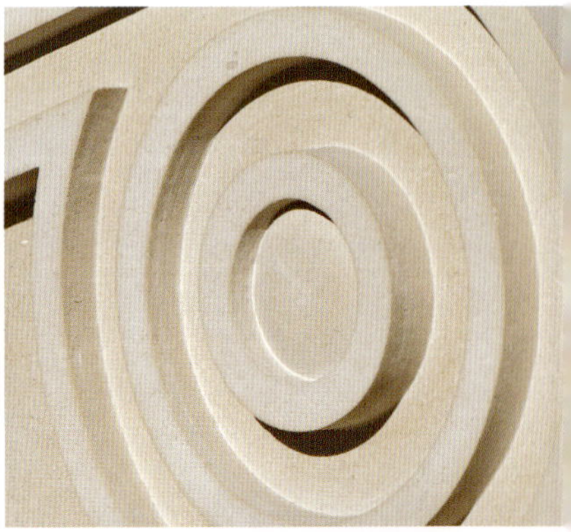
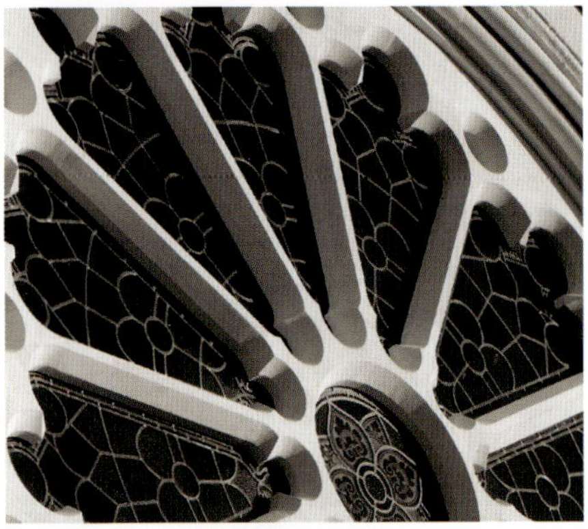
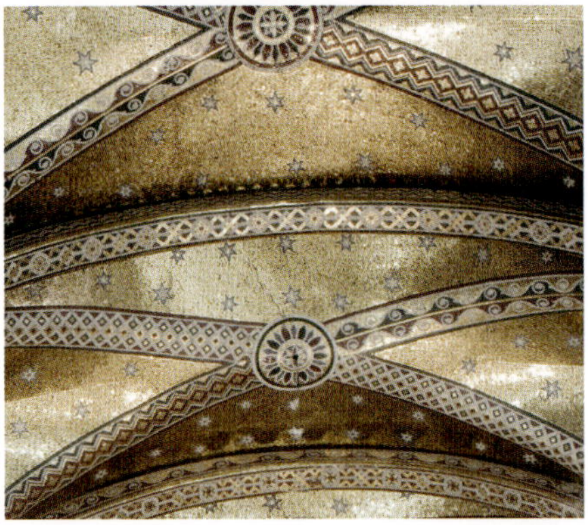
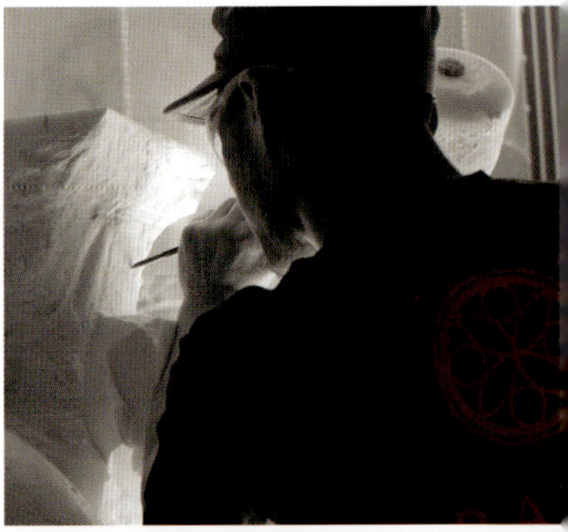
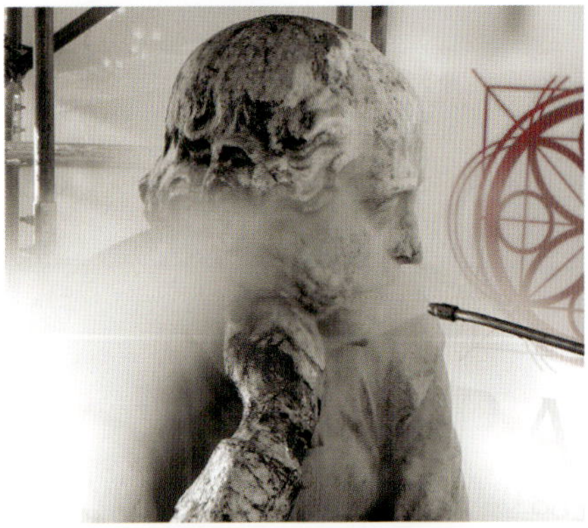
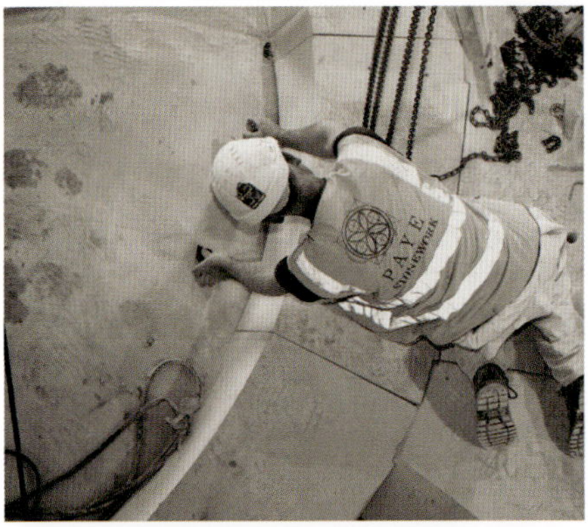
PAYE
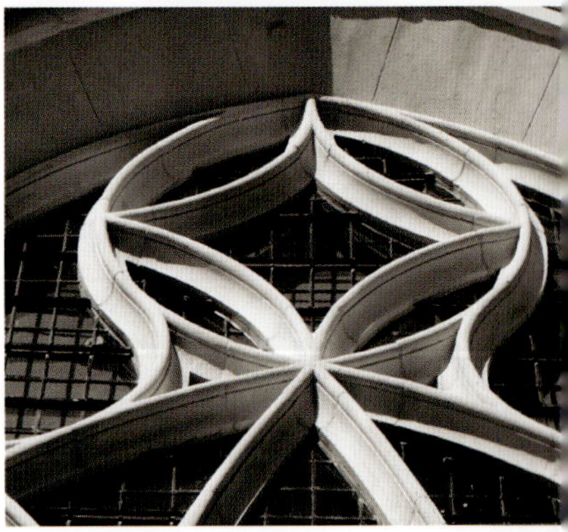
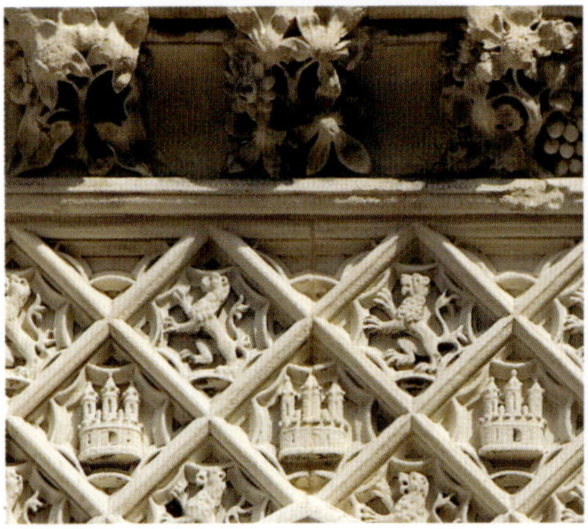
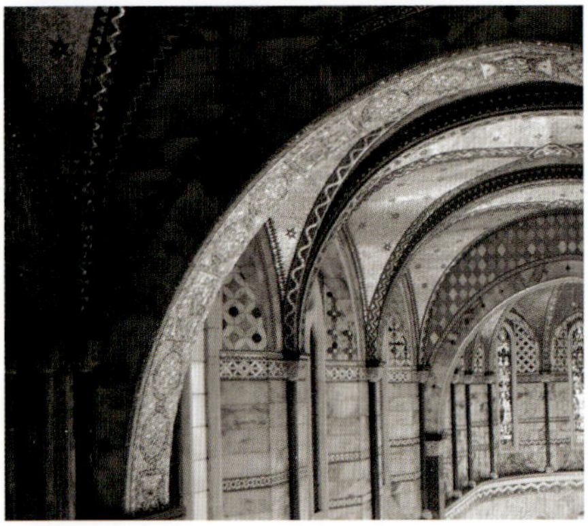
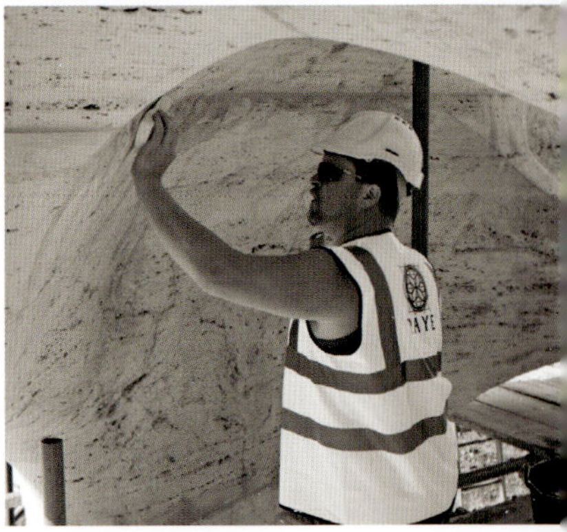
PAYE

PAYE

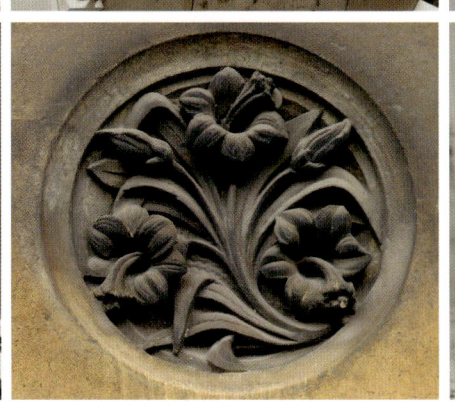

INDEX

Algae 32, 42
Aluminium oxide 39, 53
Ancaster stone 9, 156, 199
Anode 149
Anston stone 49
Artificial stone (reconsituted stone) 59, **153**, 165
Ashburton marble 12
Ashlar 93
Atmospheric pollutants 42

Basalt fibre rods 81
Bath stone 9, 11, 18, 21, 22, 55, 63, 65, 77, 93, 115, 149, 195, 196, 199, 200, 203
Bedding cills 116
Bedding planes 11, **91**
Bi-metallic reaction 141
Biocide 31, 53
Bitumen 138
Blaxter stone 199
Block bonding **109**
Board rash 60
Bonding 159
Bramley Fall stone 11, 93
Brick and tile make-up 171, 174
Brick facades 41, **155**
Brickwork 41, 42, 49, 50, 53, 59, **155**, 156, 161, 163, 195
Brick manufacture 156

Caen stone 9, 22, 73, 189, 191, 196
Calcite 39, 53
Calcium carbonate 9, 22
Candles 42
Cantilevered stairs 118
Carrara marble 12, 28, 39, 43, 47, 199
Casting concrete against masonry 116
Cast iron 133
Cast stone (see artificial, reconstituted) 153
Cathodic protection **143**, 147
Causes of failure 63, **67**
Cellulose 56
Cement mortar 18, 78, 174
Cement Mortar Repairs 18
Chains 25
Chemical cleaning 45, **49**
Chemicals 59
Chilmark stone 21
Cill 93
Cladding 115
Clay 56
Cleaning **39**
The Clean Air Act (1956) 11, 17
Clinker Concrete **22**, 168
Clipsham stone 9
Clunch stone 9
Coping 93
Common brick 159
Consolidants 28, **82**
Contamination 21
Contaminating salts 21
Cornices **32**, 109, 167, 171
Corroding metalwork/steelwork 25, 27, 28, 31, 65, 123, **135**, 137, 138, 141, 143, 147, 167
Cramps 25, 27
Craze (Crazing) 168
Crypto-florescence (also see salt crystallisation) **21**, 116
Crystallisation damage (also see salt crystallisation) 21

Decay 28
Defects **17**, (brick) 161
De-icing salts 21
Density 99, 101
Delamination 27, 28

Desalination 22
Dismantle 127
Displacement 65, 137, 143
Distribution 99, 101, 105
Domes 25
Drawings **94**
Dry pack (reconstituted stone) 153
Dry rot 174

Efflorescence (also see salt crystallisation) 28, 42, 123
Embedded metalwork (also see corroding metalwork) **25**, 27
Embedded steel (also see corroding metalwork) 22, 28, 31, 67
Engineering brick 159
Epoxy coatings 138
Erosion 9, 11, 12, 31
Extraneous metalwork 144

Facing brick 159
Faience 27, 41, 49, 50, 59, 70, 125, 127, 137, 141, 143, **165**, 167, 168, 171
Ferrous cramps 25
Final clean 60
Fine nebulous water sprays 45, **47**, 53, 59
Fixings 141
Flint 63
Fracturing 65, 143
Frost damage **31**, 161
Frost resistance 177

Gauged bricks 161
Glazed brick 59
Granites 14, 22, 42, 49, 50, 59, 70, 199
Graphite anodes 147
Grit blasting 45, **53**, 177
Grouting **109**

High pressure water lance **45**
Historic use of metals **133**
History of brickwork 155
Hoisting 101, 105
Holding exercise 143
Hopton Wood stone 9
Hot water/steam cleaning **50**, 53, 59
Hydrofluoric acid 41, 49

Ice crystals 31
Inappropriate treatments 68
Injuries 105
Installation 99, 101
Installation sequence 96

Jesmonite 32
Joint patterns 141

Kentish Ragstone 9, 65, 82, 93, 94, 181, 182, 189, 191, 196
Kentledges 105

Lichens 32, 42
Lifting equipment **105**
Limestone **9**, 18, 22, 41, 42, 47, 49, 50, 56, 59
Lime putty 56
Limewatering 82
Linseed oil 82
Lead capping 31, 34, 68, 174
Leaking rainwater pipes 56
Lewis pins 105, 106
Load path 91
Low pressure air abrasion 45, **53**, 59

Maintenance 67
Maintenance periods 77
Maintenance Strategies **141**

Mansfield Stone 199
Manufacture **156**
Marble **12**, 42, 45, 50, 59, 119
Marine environments 21, 22, 81, 137, 141
Marine salt 21
Metal stains 42, 56
Metamorphic stone 12
Method statement 99, 127
Mild steel 133
Moleanos limestone 118
Mortar 70, 73, 78, 82, 109, 163, 168, 177, 179, 181, 182, 184
Mortar repairs (also see cement mortars) **78**
Mosaic 45
Mosses 32
Movement joints **109**, 110

Nano limes 87

Oolitic limestones 9
Olivine 39
Ordinary Portland cement (OPC) **68**

Paint 82
Paint removal 174
Paraffin 82
Parker's Roman Cement 171, 177
Paving stones 11, 93
Pea-soupers, 18
Pediments **34**
Personal protection equipment (PPE) 50, **99**
Planned maintenance 63
Pointing 73
Polishing (limestone) 9
Pollution 11, 17, 18, 22, 27, 42, 168
Pollutants 22, 42
Polyurethane 82
Portland stone 9, 11, 27, 32, 37, 45, 47, 49, 53, 55, 56, 59, 60, 67, 70, 75, 93, 94, 110, 115, 116, 119, 128, 131, 133, 135, 137, 138, 143, 179, 196, 199, 200, 203
Poultices/Poulticing 22, 45, 49, **56**, 59, 116, 174
Pressurised hot water/steam 45, **50**, 53, 59, 177
Purbeck marble 12

Rainwater 27, 123
Rainwater ingress & management 27, **28**, 31, 67, **68**, 123, 135, 138, 163
Reconstituted stone (artificial stone) 59, **153**, 165
Recon stone 153
Reconstruction 127
Red Mansfield stone 28, 93, 199, 200
Red oxide paint 138
Redressing (re-working) stone **75**
Regent Street Disease (also see corroding metalwork) 135
Reigate stone 191, 196
Repairs **63**, **163**
Replacement masonry **141**
Replacing stone **75**
Repointing **70**
Restraint ties 25
Retained façade 123, 124, 125
Ribbon anodes 147
Risk assessment 99
Riven 12
Roman cement 171, 174, 177

Salts 22, 49, 161, 177
Salt contamination 22
Salt crystallisation 39, 81, 116, 161
Salt damage **21**, 22
Sandstones **11**, 41, 42, 47, 49, 50, 59, 91
Sand-blasting 39

Saturation 27
Scaffolding **101**
Scheduling **94**
Selection **93**
Shellac 82
Shelter coats 22, **81**
Shims **109**
Silica 99
Silicon 82
Silicon treatments 27
Silicosis 39, 53
Siloxanes (Silanes) 28, 82, 87
Slates **14**
Smogs 18
Sodium hydroxide 49
Soft bristle brushing 47
Soluble organic residues 31, 42, 47, 56, 116, 1[?]
Soot 39
Special Sites of Scientific Interest 32
Spiders 25
Spires 25
Sponges 45
Staining 28, 31, 32, 42, 47, 49, 116
Stainless steel 133
Stanton Moor stone 11
Steam cleaning 39
Steel frame 143, 147
Steel treatments **138**
Stitching fractures **81**
Stitching rods 177
Stone dust 177
Storing 127
Strength in compression **91**
Stucco render 32, 153, 155, **171**, 174, 177
Sulphates 22, 41, 42
Sulphate encrustation 41
Sulphuric acid 22
Surface treatments 22, **27**
Surveys **67**, 137
Swabs 45

Tobacco 42
Terra cotta (also see faience) 22, 25, 41, 49, 50, 59, 87, **165**, 167, 168, 171, 199, 200
Testing 93
Thin cladding **110**
Timber lintels 174
Tolerances **106**
Tooling marks 75
Traditional masons' water cleaning 47
Travertine marble 14, 113, 128
Treatment of steel frames (also see corroding metalwork) 138

Underclean 42
Under-firing 168
Unloading 99, 105

Varnish 82
Vegetation 31

Water ingress 31
Water traps 75
Wax 82
Wealden stone 11, 17, 25
Wet cast (reconstituted stone) 153
Weathering **17**, 75
Wrought iron 133

York stone 11, 93, 119, 199
York stone corbel 171, 174

Zinc 138
Zinc based paints 138

212

The author has made every effort to contact holders of copyright works. Anyone we have been unable to reach is invited to contact the author so that a full acknowledgement may be given in subsequent publications. For permission to reproduce the images in this book and for supplying photographs, the author thanks those listed below:

Peter Jeffree
Vivienne Freeman
Ade Groom
Angus Smith
Henry Moore's The Arch reproduced by permission of The Henry Moore Foundation

Authors Note:
Photographs are from projects undertaken within the last 20 years and represent health and safety procedures applicable at the time. They do not necessarily reflect current practices.

FSC MIX Paper from responsible sources FSC® C019400

Project management Vivienne Freeman
Inital design by Sam Pope, Twig Inc
Design development, artwork and production by Sally Eldars, **dmicreative.com**